KeySQL: The Definitive Guide

SQL for NoSQL Data

Mikhail Gilula

Andrey Belyaev

KEYARK

Contents

5 KeySQL functions and operators 162

Preface

KeySQL query language is rooted in the Key-object Data Model presented in the book *"Structured Search for Big Data: From Keywords to Key-objects"*, *Elsevier (2015),* by Mikhail Gilula. This data model is virtually the simplest possible as it is based on a single operation called *composition,* yet it is capacious enough for data of any complexity.

It continues the line of thought of Edgar F. Codd who in 1969 introduced the Relational Data Model with the goal of raising the productivity of database programming. His Turing Award lecture of 1981 was titled *"Relational Database: A Practical Foundation for Productivity".* The relational database systems with SQL query language have formed a multi-billion-dollar industry and were considered the golden standard.

One of the limitations of the relational model is the difficulty of handling complex (*non-flat*) data objects, especially those containing repeating elements. A difficulty of the same nature is the structural mismatch between the relational tables and the non-flat data objects in application programs. The need for data conversion between application programs and relational databases can degrade the overall performance of IT systems. This negative effect gained the name *impedance mismatch.*

These issues became critical in the Internet age and caused the NoSQL movement which abandoned Codd's *data independence* principle in favor of essentially storing data in files rather than in relational databases. NoSQL systems lack an SQL-like set of high-level operators and generally are not suited for ad hoc querying. They tilt the balance between productivity of programming and software performance towards the latter and create barriers for data analysts fluent in SQL but not in Java, Scala, Python, or other languages needed to speak natively to NoSQL systems.

Therefore, we thought it worthwhile to create a database language that would follow the SQL paradigm but could handle both flat and non-flat data, avoiding the impedance mismatch. Hence the key-object data model. Hence the KeySQL language, which we believe builds a practical foundation for productivity for programmers and non-programmers, such as business analysts. This time (help us Codd!) in the non-flat world.

The reader will judge if we are on track to achieving this goal.

Software downloads

The KeySQL® Scalable Platform is a project of Keyark, Inc. that includes the following components.

KeySQL Server

KeySQL Server is a parallel and scalable high-performance transactional engine supporting k-object data model and written in C++. Its transparent and replaceable data storage is currently formed by industry leading and open-source databases supporting standard SQL. The server can work with several heterogeneous databases at the same time, providing a performant scalable data layer over a plurality of SQL databases. Other types of data storage will be added in the future.

KeySQL Studio

KeySQL Studio is a web application that runs in all popular browsers. Studio provides an elegant user interface for analyst and administrator interaction with KeySQL Server. Studio's import function makes for easy loading of JSON/BSON, XML, and CSV data. There is no need to define the data structure before import. Studio can automatically recognize the schema of JSON/BSON data, including the data type, repeating values, and arrays. Upon import the structure of data is easily viewed and navigated via the Studio Explorer, a tree-like display of the data structure. Data export to JSON, XML, and CSV formats is also available.

Data Modeling Tool

The KeySQL Data Modeling Tool is a state-of-the-art web application for modeling complex, non-flat data. It is useful for: design and revision of non-flat data models; creation of KeySQL catalogs with their k-objects; documenting and sharing data models via on-screen and printed JPEG and SVG diagrams; deciphering the schema of complex data, whether it be from JSON, XML, MongoDB, or other NoSQL databases. The Data Modeling Tool can be run stand-alone but works in tandem with KeySQL Studio to deliver additional functionality.

REST API

The KeySQL REST API is a standard API that supports HTTP requests for accessing KeySQL Server. A wide range of Business Intelligence tools such as MS Power BI can directly query KeySQL Server via this API. In addition, the API can be used to rapidly move data between KeySQL Server and databases, data lakes, and proprietary systems.

KeySQL Java SDK

Java developers are provided with a Java SDK and driver fully supported by Keyark. Novice developers can compose commands in KeySQL Studio, such as "SELECT Name, Email FROM Contacts", embed this code into a single Java statement, and fetch the result set, with only a few lines of simple code.

KeySQL Python SDK

Python developers are provided with a Python SDK and driver fully supported by Keyark. The SDK is similar to the Java SDK, providing a friendly environment for novice developers as well as a sophisticated class library for more advanced developers.

This software can be downloaded via **www.keyark.com** website. The KeySQL Server, KeySQL Studio, and REST API are available within one docker image, which can be used e.g. for running the KeySQL examples presented in this book on the reader's computer of choice: Windows, Mac, or Linux one. Just copy a KeySQL code example from the book into the Studio and execute it.

Acknowledgements

It is our privilege to work in a team with our colleagues whose hard work and top-notch professionalism has made the KeySQL Platform a reality, and our thanks go to Konstantin Andreyev, Natalia Belyaeva, David Golden, Oleg Kirianov, Andris Korzans, Sergei Kosachev, Mikhail Lapshin, Victoria Pinchuk, Serguei Tarassov, and Grigory Zaigralin. Serguei Tarassov also created most Backus-Naur syntax descriptions in this book.

We are deeply indebted to Denis Sverdlov for his great input into the project.

Introduction

This guide serves as the definitive reference for understanding the KeySQL query language.

If you are familiar with the SQL language that manages data held in relational database management systems (RDBMS), then you are already familiar with KeySQL.

It has been over 50 years since the breakthrough of E.F. Codd's relational model that underlies most of today's databases. It has been nearly that long since adoption of the SQL language that provides easy access to the flat data managed in the relational model.

Big data is demanding that databases accommodate a greater variety of data, specifically non-flat data with hierarchies and repeated values commonly encountered in the likes of JSON or XML. This non-flat data does not fit comfortably within the relational model, and consequently is handled poorly and queried with great effort.

KeySQL was invented to readily manage all enterprise data, including the fast-growing volumes of non-flat data. It adheres to the SQL standard of productivity while dealing with data structures much more complex than relational tables. And it makes the powerful RDBMS functionality including joins, set operations, and aggregations, equally applicable to the non-flat data.

KeySQL data model and basic constructs 1

KeySQL is a structured query language that employs a *keyobject* (*k-object*) data model. The k-objects work as data definitions for *k-object instances*, which represent the data itself. K-object instances are formed as pairs consisting of a *key* (or *name*) and a *value* divided by a colon and enclosed in braces. The key, which is the k-object name, is a character string starting with a letter, and the value can be either *elementary* like a number, a character string, or a date, or be a set (a *composition*) of k-object instances.

Consider the following four examples of k-object instances.

`{LAST_NAME:'Johnson'}`

This is an instance of an elementary k-object `LAST_NAME` of the character type which has been assigned a value of `'Johnson'`.

`{AGE:20}`

This is an instance of an elementary k-object `AGE` of the number type which has been assigned a value of `20`.

`{PERSON:{LAST_NAME:'Doe', FIRST_NAME:'Jane'}}`

This is an instance of a *composite* k-object `PERSON` which is a set or a composition of elementary k-objects `LAST_NAME` and `FIRST_NAME`. The value of `PERSON` is a composition of instances of `LAST_NAME` and `FIRST_NAME` where `LAST_NAME` has been assigned a value of `'Doe'` and the `FIRST_NAME` has been assigned a value of `'Jane'`.

`{PEOPLE:{PERSON:{LAST_NAME:'Smith',FIRST_NAME:'John'},`
`        PERSON:{LAST_NAME:'Brown',FIRST_NAME:'Jane'}}}`

This is an instance of a *multi-composite* k-object `PEOPLE` which is a *multi-composition* of k-object `PERSON`. This means that the value of an instance of `PEOPLE` is a composition of one or more instances of `PERSON`. Here, the value of `PEOPLE` is a composition of two instances of `PERSON`.

Sets of k-object instances form *data stores*, or simply *stores*, which can be viewed as generalized analogs of SQL tables and databases at the same time. Consider an example of a data store consisting of the same four k-object instances as follows.

```
{LAST_NAME:'Johnson'},
{AGE:20},
{PERSON:{LAST_NAME:'Doe',FIRST_NAME:'Jane'}},
{PEOPLE:{PERSON:{LAST_NAME:'Smith',FIRST_NAME:'John'},
         PERSON:{LAST_NAME:'Brown',FIRST_NAME:'Jane'}}}
```

From the standpoint of mathematics, KeySQL data model is founded in *hereditarily finite sets*. Data definition and manipulation in KeySQL are based on a construct called *composition*. It is the only one needed for producing new k-objects and k-object instances from the existing ones. Essentially, the composition creates a named set from given elements, which in turn can be the named sets produced as the composition of other elements.

For comparison, SQL data model is founded in the relational algebra that comprises operations like *projection, restriction,* and *join.* KeySQL uses their generalized analogs which allow for objects of arbitrary complexity rather than flat SQL tables. KeySQL employs the same high-level data manipulation statements SQL does, i.e., INSERT, SELECT, UPDATE, and DELETE that are applied to stores rather than tables.

1.1   Bottom–up data definition: k-objects and catalogs

In KeySQL, data definition starts with creating at least one *catalog*. The catalogs are namespaces that ensure uniqueness of k-object names within each catalog. Catalogs are created using CREATE CATALOG statement as follows.

`CREATE CATALOG MY_CATALOG`

Each catalog is a set of k-objects that are defined using CREATE KEYOBJECT statement.

K-objects are defined bottom up. We start by defining three elementary k-objects: FIRST_NAME and LAST_NAME of CHAR type, and AGE of NUMBER type as follows.

```
CREATE KEYOBJECT AGE NUMBER IN CATALOG MY_CATALOG;
CREATE KEYOBJECT FIRST_NAME CHAR IN CATALOG MY_CATALOG;
CREATE KEYOBJECT LAST_NAME CHAR IN CATALOG MY_CATALOG;
```

To create a more complex k-object PERSON comprising k-objects LAST_NAME and FIRST_NAME, the composition construct {} is used as follows.

```
CREATE KEYOBJECT PERSON {LAST_NAME, FIRST_NAME} IN CATALOG MY_CATALOG
```

The keyword CATALOG in the CREATE KEYOBJECT statement is optional.

The composition construct can also be used to create a k-object PEOPLE comprising multiple (one or more) occurrences of k-object PERSON as follows.

```
CREATE KEYOBJECT PEOPLE {PERSON MULTIPLE} IN CATALOG MY_CATALOG
```

We call it *multi-composition* to distinguish between the two ways the composition is applied.

Using the composition and multi-composition, k-objects of any complexity can be created, and there are no other constructs allowed in KeySQL.

For example, since a multi-composition applies exclusively to a single elementary, composite, or multi-composite k-object, the following k-object definition is invalid.

```
CREATE KEYOBJECT INVALID {LAST_NAME, FIRST_NAME MULTIPLE} IN
CATALOG MY_CATALOG
```

Another example of an invalid k-object definition is as follows.

```
CREATE KEYOBJECT INVALID {LAST_NAME, LAST_NAME} IN CATALOG
MY_CATALOG
```

It is invalid because k-object LAST_NAME occurs more than once in the composition.

For comparison, the following chain of definitions is legitimate.

```
CREATE KEYOBJECT PERSON_1 {LAST_NAME, FIRST_NAME} IN MY_CATALOG;
CREATE KEYOBJECT PERSON_2 {LAST_NAME, FIRST_NAME} IN MY_CATALOG;
CREATE KEYOBJECT PERSON_12 {PERSON_1, PERSON_2} IN MY_CATALOG;
```

1.2   Stores: inserting data

The data in KeySQL is laid out as k-object instances that are inserted into stores. Each store is based on a single catalog specified in the store definition and can generally hold instances of any k-object from the catalog. Therefore, KeySQL stores can hold data records with different logical structures unlike SQL tables where all rows must consist of the same set of columns.

The stores are defined using CREATE STORE statement as follows.

```
CREATE STORE MY_STORE FOR CATALOG MY_CATALOG
```

The keyword CATALOG in the CREATE STORE statement is optional.

K-object instances are inserted into stores using INSERT INSTANCES (or INSERT SELECT) statement, for example, as follows.

```
INSERT INTO MY_STORE INSTANCES
{LAST_NAME:'Johnson'},
{AGE:20},
{PERSON:{LAST_NAME:'Doe',FIRST_NAME:'Jane'}},
{PEOPLE:{PERSON:{LAST_NAME:'Smith',FIRST_NAME:'John'},
        PERSON:{LAST_NAME:'Brown',FIRST_NAME:'Jane'}}};
```

The k-object instances inserted into a store using the INSERT statement are called *host k-object instances,* or simply *host instances* to distinguish them from the instances they *host* (contain) as their elements or the elements of the elements, and so forth. In the example above, the host instance of k-object PERSON hosts the instances of k-object LAST_NAME and k-object FIRST_NAME. And the host instance of k-object PEOPLE hosts two instances of PERSON, two instances of LAST_NAME, and two instances of FIRST_NAME.

> Note that k-object definitions in a catalog are shared across all stores defined for the catalog.

1.3   Simplest queries

In KeySQL, queries are addressed to stores, and a query result is a store consisting of k-object instances specified by the query. In SQL, queries are addressed to tables, and a query result is a table with columns and rows specified by the query. Unlike a resulting SQL table, a resulting KeySQL store can contain instances of different k-objects, that is, the instances in the resulting store can have different structures.

KeySQL queries often look like SQL queries. For example, the simplest queries are as follows.

```
SELECT *
FROM MY_STORE
```

The result is always the original store. It consists of all host instances comprising the store at the time of the query execution. In our case, these are the instances inserted into MY_STORE in the last example of the previous section as follows.

```
{LAST_NAME:'Johnson'},
{AGE:20},
{PERSON:{LAST_NAME:'Doe',FIRST_NAME:'Jane'}},
{PEOPLE:{PERSON:{LAST_NAME:'Smith',FIRST_NAME:'John'},
         PERSON:{LAST_NAME:'Brown',FIRST_NAME:'Jane'}}}
```

> The order of elements within a returned k-object instance matches the order in the k-object definition (CREATE/ALTER statement) unless the SELECT statement specifies another order.

1.4   Projection on k-object

In KeySQL, the SELECT clause specifies the structure of k-object instances in the resulting store. In SQL, it specifies the columns of the resulting table. Like in the relational model and SQL, the operation behind that in KeySQL is called *projection*.

In SQL, it projects the original table on one or more columns listed in the SELECT clause which become the columns of the resulting table. In KeySQL, it projects each host instance on one or more k-objects listed in the SELECT clause. Since a k-object structure can be complex, traditional SQL projection needs some generalization.

Consider the simplest case of projecting our exemplary MY_STORE on a single k-object LAST_NAME using the following query.

```
SELECT LAST_NAME
FROM MY_STORE
```

To form the resulting store, we project each of four host instances in MY_STORE on LAST_NAME.

The result of projecting the first and the third host instances is intuitive. Each of them returns an instance of LAST_NAME as follows.

```
{LAST_NAME:'Johnson'},
{LAST_NAME:'Doe'},
...
```

The second host instance does not reference LAST_NAME. Its projection on this k-object is empty and does not contribute to the result. This does not happen in SQL where each row of a table consists of the same columns and brings a row into the resulting table.

The fourth instance takes us further from SQL. It has two occurrences of LAST_NAME. Therefore, we need to combine the two LAST_NAME instances into a single resulting instance of a k-object being a multi-composition of LAST_NAME, like the one created by the following statement.

```
CREATE KEYOBJECT LAST_NAMES {LAST_NAME MULTIPLE} IN MY_CATALOG
```

But we do not have it in our catalog. It could be automatically added to the catalog and used in the query results. However, in order not to fill catalogs with k-objects that may be used just once, there is another option.

For ad hoc use in query results KeySQL employs a naming convention showing the multiple of what k-object we are dealing with. It adds the hash sign "#" in the beginning of a k-object name, and the user-created k-objects in KeySQL cannot start with the hash sign. This technique yields the projection of our fourth host instance on LAST_NAME as follows.

```
{#LAST_NAME: {LAST_NAME:'Smith', LAST_NAME:'Brown'}}
```

The complete result of the query is shown below.

```
{LAST_NAME:'Johnson'},
{LAST_NAME:'Doe'},
{#LAST_NAME: {LAST_NAME:'Smith',LAST_NAME:'Brown'}}
```

Consider now a case of projecting on a composite k-object as follows:

```
SELECT PERSON
FROM MY_STORE;
```

Though `MY_CATALOG` includes k-object `PEOPLE`, KeySQL still uses the default naming convention unless it is instructed otherwise, and the result of the query is as follows.

```
{PERSON {FIRST_NAME : 'Jane',LAST_NAME:'Doe'}},
{#PERSON: {PERSON: {FIRST_NAME:'John',LAST_NAME:'Smith'},
          PERSON: {FIRST_NAME:'Jane',LAST_NAME:'Brown'}}}
```

The composite and elementary k-objects are treated the same way. Queries are formulated using the names of k-objects independently of their internal structure (i.e., the k-objects they contain) however complex it may be. Should the k-object's internal structure vary, the query will remain the same.

For comparison, in SQL, entities like `PERSON` and `PEOPLE` must be represented by one or more tables, and the table names cannot be used in the SELECT lists. Therefore, SQL imposes the use of specially generated identifiers that link up the tables; and SQL queries must reference columns like `PERSON_ID` instead of the entity `PERSON` itself.

1.5 Projection on set of k-objects

Consider the following query:

```
SELECT LAST_NAME, FIRST_NAME
FROM MY_STORE
```

In SQL, it returns a two-column table with the original columns `LAST_NAME` and `FIRST_NAME`. In KeySQL, the result comprises projections of each host instance on the set of two k-objects: `LAST_NAME` and `FIRST_NAME`.

We already know how the projection on a single k-object works. A *projection on a set of k-objects* is formed as the composition of projections on each k-object of the set. And all k-objects in the SELECT list must be referenced by the host instance, otherwise its projection is empty.

The composition is formed as an instance of an ad hoc k-object that is assigned a reserved name `RESULT`. This name is not allowed in catalogs. The components of `RESULT` are either k-objects from the `SELECT` clause or their multiples which are named using the hash symbol introduced in the previous section.

The result of our query is formed as follows. The first host instance references `LAST_NAME` but not `FIRST_NAME`, and therefore its projection is empty. The second host instance also produces an empty projection. The third one references both `LAST_NAME` and `FIRST_NAME` and returns the following.

```
{RESULT: {LAST_NAME:'Doe', FIRST_NAME:'Jane'}}
```

The projection of the fourth host instance contains multiples, so we see `#LAST_NAME` and `#FIRST_NAME` as follows.

```
{RESULT: {#LAST_NAME: {LAST_NAME:'Smith', LAST_NAME:'Brown'},
          #FIRST_NAME: {FIRST_NAME:'John', FIRST_NAME:'Jane'}}}
```

Accordingly, the complete query result contains two instances as follows.

```
{RESULT: {LAST_NAME:'Doe', FIRST_NAME:'Jane'}},
{RESULT: {#LAST_NAME: {LAST_NAME:'Smith', LAST_NAME:'Brown'},
          #FIRST_NAME: {FIRST_NAME:'John', FIRST_NAME:'Jane'}}}
```

Notice that the projection of `PEOPLE` relates the sets of `LAST_NAME` and `FIRST_NAME` values occurring in the host instance but not the correspondence between them. Given the results it is impossible to tell which of the first names relates to the last name `'Brown'`, for example. So, the query really means: "*Get all last names and all first names from each instance in the store.*"

There are of course means in KeySQL to get the pairs of last and first names as they occur in the host instances, and we shall see those queries in the following sections. However, the query at hand does not reflect any stipulation that the last and the first name must be returned in pairs.

The SQL intuition does not work well when dealing with multiple values. However, for the flat data, the KeySQL projection works the same way as the SQL projection does. For example, consider the projection of the flat `PERSON` host instance above.

The projection results can reference ad hoc k-objects. However, users may specify the catalog k-objects to be used in query responses. The way it works is illustrated in the next sections.

The topic of ad hoc k-objects will be revisited further in this chapter where this functionality will be extended to simplify ad hoc querying by employing *user-defined ad hoc k-objects*.

> Projection is independently applied to each host instance in a store and at a maximum produces one resulting instance from each host instance. Projection output may reference special standard ad hoc k-objects if the

1.6   Renaming projection results: operator TO

Query results in the previous sections reference the ad hoc k-objects. This has its analogy in SQL where some column names in the resulting tables are generated automatically if users do not explicitly name those columns. A way to specify a column name in SQL is to put the AS keyword between a select list item and its desired name.

KeySQL employs two analogs of the SQL AS clause. Those are operators TO and AS. The first one does pure renaming while the second one can also restructure the default projection output. Therefore, though TO and AS operators can sometimes produce the same results, they work differently and hold their specific positions in the order of KeySQL query processing.

This section introduces operator TO. The AS operator is considered in the next section.

Let us return to our first projection query:

```
SELECT LAST_NAME
FROM MY_STORE
```

As we know, it produces the following resulting data set.

```
{LAST_NAME:'Johnson'},
{LAST_NAME:'Doe'},
{#LAST_NAME: {LAST_NAME:'Smith', LAST_NAME:'Brown'}}
```

To avoid getting the generated name #LAST_NAME denoting the multiple of the LAST_NAME we shall create a k-object LAST_NAMES as follows.

```
CREATE KEYOBJECT LAST_NAMES {LAST_NAME MULTIPLE} IN MY_CATALOG;
```

Now we can use it in the query as follows.

```
SELECT { LAST_NAME } TO LAST_NAMES
FROM MY_STORE
```

Here, the braces around LAST_NAME mean that, for each host instance, all instances of k-object LAST_NAME from its projection on this k-object must be combined into an instance of k-object `LAST_NAMES`. The catalog k-object `LAST_NAMES` not only replaces the ad hoc k-object `#LAST_NAME` but must also be used even in the case when a host instance has only one occurrence of the `LAST_NAME`, which otherwise would form its input into the resulting store.

And we get the following resulting store.

```
{LAST_NAMES: {LAST_NAME:'Johnson'}},
{LAST_NAMES: {LAST_NAME:'Doe'}},
{LAST_NAMES: {LAST_NAME:'Smith', LAST_NAME:'Brown'}}
```

Notice that the structure of the resulting store is now homogeneous. All host instances are the instances of `LAST_NAMES` even if hosting just a single occurrence of `LAST_NAME`. This is different from the ad hoc projection results which can return a mixture of `#LAST_NAME` and `LAST_NAME`.

Consider now our second projection query, which was as follows.

```
SELECT PERSON
FROM MY_STORE
```

Its familiar result is below.

```
{PERSON: {FIRST_NAME:'Jane', LAST_NAME:'Doe'}},
{#PERSON: {PERSON: {FIRST_NAME:'John', LAST_NAME:'Smith'},
           PERSON: {FIRST_NAME:'Jane', LAST_NAME:'Brown'}}}
```

Here, to get rid of the ad hoc k-object `#PERSON` we can use the already existing k-object `PEOPLE`, and the query looks as follows.

```
SELECT {PERSON} TO PEOPLE
FROM MY_STORE
```

It produces the following resulting store.

```
{PEOPLE: {PERSON: {FIRST_NAME:'Jane', LAST_NAME:'Doe'}}},
{PEOPLE: {PERSON: {FIRST_NAME:'John', LAST_NAME:'Smith'},
          PERSON: {FIRST_NAME:'Jane', LAST_NAME:'Brown'}}}
```

Let us now work on our third projection query in the same manner. It looked as follows.

```
SELECT LAST_NAME, FIRST_NAME
FROM MY_STORE
```

And produced the following resulting data set.

```
{RESULT: {LAST_NAME:'Doe', FIRST_NAME:'Jane'}},
{RESULT: {#LAST_NAME: {LAST_NAME:'Smith', LAST_NAME:'Brown'},
          #FIRST_NAME: {FIRST_NAME:'John', FIRST_NAME:'Jane'}}}
```

To "fix" it, we need not only the multiple of the just defined LAST_NAME but also the multiple of the FIRST_NAME that is analogously defined as follows.

```
CREATE KEYOBJECT FIRST_NAMES {FIRST_NAME MULTIPLE} IN CATALOG
MY_CATALOG
```

We also need a new composite k-object that can accommodate LAST_NAMES and FIRST_NAMES as follows.

```
CREATE KEYOBJECT LAST_FIRST_NAMES {LAST_NAMES, FIRST_NAMES} IN
CATALOG MY_CATALOG
```

Now we are ready to rewrite the query using the renaming operator TO as follows.

```
SELECT {{LAST_NAME} TO LAST_NAMES, {FIRST_NAME} TO FIRST_NAMES} TO
LAST_FIRST_NAMES
FROM MY_STORE
```

The resulting store is as follows.

```
{LAST_FIRST_NAMES: {LAST_NAMES: {LAST_NAME:'Doe'},
                    FIRST_NAMES: {FIRST_NAME:'Jane'}}},
{LAST_FIRST_NAMES: {LAST_NAMES: {LAST_NAME:'Smith',
LAST_NAME:'Brown'},
                    FIRST_NAMES: {FIRST_NAME:'John',
FIRST_NAME:'Jane'}}}
```

> Operator TO is applied to the projection results. It does not change the number or the structure of k-object instances in the output but just renames them using the catalog k-objects or the user-defined ad hoc k-objects specified in the SELECT clause.

1.7   Restructuring projection results: operator AS

With or without operator TO, projection queries always produce no more than one resulting instance from each of the host instances. However, there are use cases where we would like to have more than one resulting instance generated from the host instances. For example, this could be the case when a host instance contains multiple instances of some k-objects which we would like to lay out as separate instances in the query results.

Operator AS can also be used to "flatten" k-object instances containing the multiples. This allows feeding KeySQL query results to BI tools, or machine learning software, or converting them into spreadsheets. An example of this use case is considered in a separate section of this guide.

Let us return to our first projection query:

```
SELECT LAST_NAME
FROM MY_STORE
```

As we know it produces the following resulting data set.

```
{LAST_NAME:'Johnson'},
{LAST_NAME:'Doe'},
{#LAST_NAME: {LAST_NAME:'Smith', LAST_NAME:'Brown'}}
```

Suppose we want the result to consist of just the instances of LAST_NAME irrespective of the host instances they originate from. It can be done with the help of AS operator as follows.

```
SELECT LAST_NAME AS LAST_NAME
FROM MY_STORE
```

Now the resulting store looks as follows.

```
{LAST_NAME:'Johnson'},
{LAST_NAME:'Doe'},
{LAST_NAME:'Smith'},
{LAST_NAME:'Brown'}
```

In this layout the information that Smith and Brown originate from the same host instance of PEOPLE is lost.

For the current state of MY_STORE, the next query with AS operator produces the same result as the one from the previous section that used the TO operator.

```
SELECT {LAST_NAME} AS LAST_NAMES
FROM MY_STORE
```

Namely, the resulting store is as follows.

```
{LAST_NAMES: {LAST_NAME:'Johnson'}},
{LAST_NAMES: {LAST_NAME:'Doe'}},
{LAST_NAMES: {LAST_NAME:'Smith', LAST_NAME:'Brown'}}
```

Though, for `MY_STORE` at hand, the results of applying TO and AS are the same, they can be different for other stores.

Let us now use AS operator to modify our second projection query as follows.

```
SELECT {LAST_NAME, FIRST_NAME} AS PERSON
FROM MY_STORE
```

Here we want the result to be presented via k-object `PERSON` from the catalog rather than via the ad hoc k-object RESULT employed by the simple projection. The result is as follows.

```
{PERSON: {FIRST_NAME:'Jane', LAST_NAME:'Doe'}},
{PERSON: {FIRST_NAME:'John', LAST_NAME:'Smith'}},
{PERSON: {FIRST_NAME:'Jane', LAST_NAME:'Brown'}}
```

Like in the first modified query, the information that `Smith` and `Brown` originate from the same host instance is lost.

Note that using TO operator instead of AS here produces a run-time error as follows.

```
SELECT {LAST_NAME, FIRST_NAME} TO PERSON
FROM MY_STORE
```

> Error: invalid type for PERSON

This is because the query attempts the renaming of the projection to k-object `PERSON` which is a flat pair of `LAST_NAME` and `FIRST_NAME`. But in fact, as we have seen, the projection produces a k-object instance consisting of multiples of the `LAST_NAME` and `FIRST_NAME`. Those multiples do not fit into the flat k-object `PERSON`, hence the error. If the projection results would not contain the multiples (which could be the case for some store), no error would occur.

Let us now use a couple of AS operators in a query as follows.

```
SELECT {{LAST_NAME,FIRST_NAME} AS PERSON} AS PEOPLE
FROM MY_STORE
```

The resulting store is as follows.

```
{PEOPLE: {PERSON: {LAST_NAME:'Doe', FIRST_NAME:'Jane'}}},
{PEOPLE: {PERSON: {LAST_NAME:'Smith', FIRST_NAME:'John'}},
        {PERSON: {LAST_NAME:'Brown', FIRST_NAME:'Jane'}}}
```

The query result is the same as of the following query with renaming from the previous section.

```
SELECT {PERSON} TO PEOPLE
FROM MY_STORE
```

And of course, the following query produces the same result:

```
SELECT {PERSON} AS PEOPLE
FROM MY_STORE
```

In the next example both TO and AS work the same way as the ad hoc projection does.

```
SELECT LAST_NAME AS LAST_NAME, FIRST_NAME TO FIRST_NAME
FROM MY_STORE
```

They just "rename" the (elementary) k-objects listed in the SELECT clause to their own names.

```
{RESULT: {LAST_NAME:'Doe', FIRST_NAME:'Jane'}},
{RESULT: {#LAST_NAME: {LAST_NAME:'Smith', LAST_NAME: Brown'},
        #FIRST_NAME: {FIRST_NAME:'John', FIRST_NAME:'Jane'}}}
```

> Operator AS is independently applied to each host instance in a store and allows replacing it in the query results with one or more instances of a catalog k-object or a user-defined ad hoc k-object if it is applied in the following context:
> SELECT <valid expression> AS <k-object name>

1.8   Operator GROUP BY

KeySQL operators GROUP BY and DISTINCT provide another extension of SQL functionality. Like the AS operator they can restructure the query results but work "per store" rather than "per host instance". In other words, the output is generated after all host instances in a store are processed.

For the data in our store, the following query with GROUP BY clause produces the same result as the first query with AS operator from the previous section.

```
SELECT LAST_NAME
FROM MY_STORE
GROUP BY LAST_NAME
```

Namely, we get:

```
{LAST_NAME:'Johnson'},
{LAST_NAME:'Doe'},
{LAST_NAME:'Smith'},
{LAST_NAME:'Brown'}
```

However, if there would be any duplicate last names, they would be eliminated by the GROUP BY query but preserved by the AS query.

In standard SQL, the GROUP BY clause must list all columns of the SELECT clause that are not referenced in aggregate functions. The GROUP BY operator of KeySQL is more flexible. Consider the following query example.

```
SELECT FIRST_NAME, LAST_NAME
FROM MY_STORE
GROUP BY FIRST_NAME
```

The result is as follows.

```
{RESULT: {FIRST_NAME:'John', LAST_NAME:'Smith'}},
{RESULT: {FIRST_NAME:'Jane',
          #LAST_NAME: {LAST_NAME:'Brown', LAST_NAME:'Doe'}}}
```

GROUP BY can also be used the same way as in SQL, for example as follows.

```
SELECT FIRST_NAME, COUNT(LAST_NAME)
FROM MY_STORE
GROUP BY FIRST_NAME
```

This query produces the following output.

```
{RESULT: {FIRST_NAME:'John', COUNT_LAST_NAME:1}},
{RESULT: {FIRST_NAME:'Jane', COUNT_LAST_NAME:2}}
```

It returns the ad hoc k-object COUNT_LAST_NAME and we can rename it provided that the catalog has the needed k-object. For example, let us create k-object QUANTITY as follows.

```
CREATE KEYOBJECT QUANTITY NUMBER IN CATALOG MY_CATALOG
```

Now we can use it in the following query.

```
SELECT FIRST_NAME, COUNT(LAST_NAME) TO QUANTITY
FROM MY_STORE
GROUP BY FIRST_NAME
```

The result is shown below.

```
{RESULT: {FIRST_NAME:'John', QUANTITY:1}},
{RESULT: {FIRST_NAME:'Jane', QUANTITY:2}}
```

Consider the next two queries illustrating how GROUP BY deals with more complex k-objects. The first one is using the already introduced k-object LAST_NAMES as follows.

```
SELECT FIRST_NAME, {LAST_NAME} TO LAST_NAMES
FROM MY_STORE
GROUP BY FIRST_NAME
```

It returns the following.

```
{RESULT: {FIRST_NAME:'John',
          {LAST_NAMES: {LAST_NAME:'Smith'}}},
{RESULT: {FIRST_NAME:'Jane',
          {LAST_NAMES: {LAST_NAME:'Brown', LAST_NAME:'Doe'}}}
```

The second one groups PERSON by FIRST_NAME using k-object PEOPLE as follows.

```
SELECT FIRST_NAME, {PERSON} TO PEOPLE
FROM MY_STORE
GROUP BY FIRST_NAME
```

It produces the following result.

```
{RESULT:{FIRST_NAME:'John',
         {PEOPLE: {PERSON:{LAST_NAME:'Smith',FIRST_NAME:'John'}}}},
{RESULT:{FIRST_NAME:'Jane',
         {PEOPLE: {PERSON:{LAST_NAME:'Brown',FIRST_NAME:'Jane'},
                   PERSON:{LAST_NAME:'Doe',FIRST_NAME:'Jane'}}}}
```

To completely replace the ad hoc k-objects in the result, let us create k-object FIRST_NAME_PEOPLE as follows.

```
CREATE KEYOBJECT FIRST_NAME_PEOPLE {FIRST_NAME, PEOPLE} IN MY_CATALOG
```

Now we can use it in our query as follows.

```
SELECT {FIRST_NAME, {PERSON} TO PEOPLE} TO FIRST_NAME_PEOPLE
FROM MY_STORE
GROUP BY FIRST_NAME
```

And we get the following resulting store.

```
{FIRST_NAME_PEOPLE:{FIRST_NAME:'John',{PEOPLE:
                   {PERSON:{LAST_NAME:'Smith',FIRST_NAME:'John'}}}},
{FIRST_NAME_PEOPLE:{FIRST_NAME:'Jane',{PEOPLE:
                   {PERSON:{LAST_NAME:'Brown',FIRST_NAME:'Jane'},
PERSON:{LAST_NAME:'Doe',FIRST_NAME:'Jane'}}}}
```

1.9   Operator DISTINCT

In SQL, operator DISTINCT can always be replaced by operator GROUP BY producing the same results. It is often used to eliminate duplicates when no aggregate function is referenced in the SELECT clause. It is mostly the case for KeySQL as well, with some exceptions.

Consider the following query which is invalid in SQL but is valid in KeySQL.

```
SELECT DISTINCT LAST_NAME, DISTINCT FIRST_NAME
FROM MY_STORE
```

Here we see how the KeySQL generalizes the SQL DISTINCT functionality to cover multiple "columns" at once, and the result is as follows.

```
{RESULT:{#LAST_NAME: {LAST_NAME:'Smith', LAST_NAME:'Doe',
LAST_NAME:'Brown'},
         #FIRST_NAME: {FIRST_NAME:'John', FIRST_NAME:'Jane'}}}
```

Each k-object in the SELECT list must be either prefixed by DISTINCT or referenced in the GROUP BY clause as follows.

```
SELECT LAST_NAME, DISTINCT FIRST_NAME
FROM MY_STORE
GROUP BY LAST_NAME
```

This query produces the following result.

```
{RESULT: {LAST_NAME:'Brown', FIRST_NAME:'Jane'}},
{RESULT: {LAST_NAME:'Doe', FIRST_NAME:'Jane'}},
```

```
{RESULT: {LAST_NAME:'Smith', FIRST_NAME: 'John'}}
```

Here is an example without GROUP BY, and where only one of two SELECT list k-objects is prefixed by DISTINCT.

```
SELECT LAST_NAME, DISTINCT FIRST_NAME
FROM MY_STORE
```

It generates the following error message.

> Error: GROUP BY required.

To provide a more meaningful example of using GROUP BY and DISTINCT in a query, let us create a new store as follows.

```
CREATE STORE MY_STORE2 FOR CATALOG MY_CATALOG;
```

We populate it as follows.

```
INSERT INTO MY_STORE2 INSTANCES
{PERSON: {FIRST_NAME:'Jane', LAST_NAME:'Smith'}},
{PEOPLE: {PERSON: {FIRST_NAME:'Jane', LAST_NAME:'Smith'},
          PERSON: {FIRST_NAME:'Jane', LAST_NAME:'Smith'}}}
```

Consider now the following query.

```
SELECT FIRST_NAME, LAST_NAME
FROM MY_STORE2
GROUP BY FIRST_NAME
```

It produces the result as follows.

```
{RESULT: {FIRST_NAME:'Jane',
#LAST_NAME:{LAST_NAME:'Smith',LAST_NAME:'Smith',LAST_NAME:'Smith'}}}
```

To eliminate the duplicates of LAST_NAME we can use DISTINCT as follows.

```
SELECT FIRST_NAME, DISTINCT LAST_NAME
FROM MY_STORE2
GROUP BY FIRST_NAME
```

Now the result is as follows.

```
{RESULT: {FIRST_NAME:'Jane', LAST_NAME:'Smith'}}
```

The DISTINCT operator can be used for k-objects of any complexity, just like its GROUP BY counterpart. For example, the query

```
SELECT DISTINCT PERSON
FROM MY_STORE
```

returns:

```
{PERSON: {FIRST_NAME:'Jane', LAST_NAME:'Doe'}},
{PERSON: {FIRST_NAME:'John', LAST_NAME:'Smith'}},
{PERSON: {FIRST_NAME:'Jane', LAST_NAME:'Brown'}}
```

> Operators GROUP BY and DISTINCT are applied to a whole store rather than to each host instance independently. GROUP BY can group any k-objects by other k-objects. DISTINCT can be applied to each of several k-objects simultaneously. GROUP BY and DISTINCT can be used in the same query.

1.10   Use of GROUP BY and DISTINCT with TO and AS

> The reader can skip this section in the first reading as it represents a deeper dive into KeySQL semantics.

Generally, GROUP BY and DISTINCT queries can contain both the TO and AS operators. To understand the query results, it is important to consider the order of operators.

The order is as follows.

1. AS
2. GROUP BY
3. TO

For example, consider the following query, which produces an error.

```
SELECT {FIRST_NAME, LAST_NAME} AS PERSON
FROM MY_STORE
GROUP BY LAST_NAME
```

> Error: keyobject LAST_NAME not found in SELECT list.

To understand why, let us first consider the query without GROUP BY as follows.

```
SELECT {FIRST_NAME, LAST_NAME} AS PERSON
FROM MY_STORE
```

It returns the following result.

```
{PERSON: {FIRST_NAME:'Jane', LAST_NAME:'Doe'}},
{PERSON: {FIRST_NAME:'John', LAST_NAME:'Smith'}},
{PERSON: {FIRST_NAME:'Jane', LAST_NAME:'Brown'}}
```

Here, after applying AS operator, the connection between PERSON and LAST_NAME from the original host-instances is lost. It is impossible to tell which pairs of the first and last names belong to the same host instance.

To compare, consider the same query with AS replaced by TO.

```
SELECT {FIRST_NAME, LAST_NAME} TO PERSON
FROM MY_STORE
GROUP BY LAST_NAME
```

This query runs without errors and produces the following reply.

```
{PERSON: {FIRST_NAME:'Jane', LAST_NAME:'Doe'}},
{PERSON: {FIRST_NAME:'John', LAST_NAME:'Smith'}},
{PERSON: {FIRST_NAME:'Jane', LAST_NAME:'Brown'}}
```

To understand how the result is formed, remember that GROUP BY is applied before the TO operator is applied.

The same query without GROUP BY will produce an error that we already know, as follows.

```
SELECT {FIRST_NAME, LAST_NAME} TO PERSON
FROM MY_STORE
```

> Error: invalid type for PERSON.

And this is because now there is no GROUP BY that creates the flat pairs of the first and last names from the original host-instances which could fit into PERSON k-object. And now the TO operator is applied to the original store rather than to the one created by the GROUP BY clause.

Let us consider the use of DISTINCT operator in combination with operators TO and AS.

Generally, the use of DISTINCT with TO operator is limited to cases when operator TO renames a single k-object into a single k-object. In these cases, operator TO can always be transparently replaced by operator AS irrespectively of the presence of DISTINCT or GROUP BY for that matter, like in the following query.

```
SELECT DISTINCT FIRST_NAME TO LAST_NAME
FROM MY_STORE
```

Which produces the following result.

```
{LAST_NAME:'John'},
{LAST_NAME:'Jane'}
```

To present a less trivial example of combining DISTINCT, TO, and GROUP BY, let us create a new k-object FIRST_NAME_LAST_NAMES as follows.

```
CREATE KEYOBJECT FIRST_NAME_LAST_NAMES {FIRST_NAME, LAST_NAMES} IN
MY_CATALOG
```

Now consider the following query:

```
SELECT {FIRST_NAME, {DISTINCT LAST_NAME} TO LAST_NAMES} TO
FIRST_NAME_LAST_NAMES
FROM MY_STORE
GROUP BY FIRST_NAME
```

which returns the following result:

```
{FIRST_NAME_LAST_NAMES:
   {FIRST_NAME: 'John', LAST_NAMES : {LAST_NAME : 'Smith'}}},
{FIRST_NAME_LAST_NAMES:
   {FIRST_NAME:'Jane', LAST_NAMES: {LAST_NAME:'Brown',
LAST_NAME:'Doe'}}}
```

However, the query:

```
SELECT DISTINCT {FIRST_NAME, LAST_NAME} TO PERSON
FROM MY_STORE
```

will result in error as follows.

> Error: invalid usage of operator DISTINCT for keyobject PERSON

To double check, consider the analog of this query with GROUP BY as follows.

```
SELECT {FIRST_NAME, LAST_NAME} TO PERSON
FROM MY_STORE
```

```
GROUP BY PERSON
```

As expected, it produces albeit a more verbose error message as follows.

> Error: keyobject PERSON is the result of operator TO, so it
> cannot be used in GROUP BY

It reminds us that TO is executed after GROUP BY.

Using the operator AS instead of TO:

```
SELECT DISTINCT {FIRST_NAME, LAST_NAME} AS PERSON
FROM MY_STORE
```

works without errors and yields the following result:

```
{PERSON: {FIRST_NAME:'Jane', LAST_NAME:'Brown'}},
{PERSON: {FIRST_NAME:'Jane', LAST_NAME:'Doe'}},
{PERSON: {FIRST_NAME:'John', LAST_NAME:'Smith'}}
```

1.11   Query scope

The queries we have seen in the previous sections are applied to all host instances in a store.

The concept of a *query scope* allows to target exclusively the instances of the specified host k-objects in the store.

Consider a query that ads a query scope PEOPLE to our first projection query as follows. Note how the scope appears in parenthesis following the store name.

```
SELECT LAST_NAME
FROM MY_STORE(PEOPLE)
```

It instructs KeySQL server to look only into the host instances of k-object PEOPLE. This happens to be the fourth host instance in our simple store, and it forms the result below.

```
{#LAST_NAME:{LAST_NAME:'Smith', LAST_NAME:'Brown'}}
```

In the next query, the scope allows to address the instances of two of host k-objects as follows.

```
SELECT LAST_NAME
FROM MY_STORE(PEOPLE,PERSON)
```

It produces the following reply.

```
{LAST_NAME:'Doe'},
{#LAST_NAME:{LAST_NAME:'Smith', LAST_NAME:'Brown'}},
```

The next query returns the complete instances of the host k-object `PERSON` in scope.

```
SELECT *
FROM MY_STORE(PERSON)
```

It brings the following result:

```
{PERSON: {LAST_NAME:'Doe', FIRST_NAME:'Jane'}}
```

1.12   Qualifiers

To address an instance of a particular k-object within a store, a *qualifier* can be used. For example, as follows.

```
SELECT PEOPLE.LAST_NAME
FROM MY_STORE
```

A complete qualifier can also be used as follows.

```
SELECT PEOPLE.PERSON.LAST_NAME
FROM MY_STORE
```

If the shorter qualifier is unambiguous, it works the same way.

However, there is a difference relative to the following query.

```
SELECT PERSON.LAST_NAME
FROM MY_STORE
```

It targets both the host instance of `PEOPLE`, and the host instance of `PERSON`, and produces the following result.

```
{LAST_NAME:'Doe'},
{#LAST_NAME: {LAST_NAME:'Smith', LAST_NAME:'Brown'}}
```

1.13   WHERE clause

In SQL, the WHERE clause restricts the set of rows of the source table having their input into the resulting table. In KeySQL, WHERE clause plays the same role. It restricts the set of all host instances in a store to those that satisfy the WHERE clause condition and are considered for producing the result. Like the projection, the *restriction* needs some generalization to accommodate arbitrarily complex data objects of KeySQL.

Billiard data store

To introduce the WHERE clause functionality we use a simple data set of Billiard Store. It is meant to represent data about two ball sets comprising colored and numbered balls of a certain diameter and quantity suited for the Carom and Pool billiard games. The Carom uses just three unnumbered balls, and there are 15 numbered balls in the Pool ball set.

First, we create a catalog `BLRD_CATALOG` and the required k-objects as follows.

> Note the use of two dashes for the inline comments in KeySQL.

```
CREATE CATALOG BLRD_CATALOG;
-- Name of the game
CREATE KEYOBJECT GAME CHAR IN CATALOG BLRD_CATALOG;
-- Ball diameter in millimeters
CREATE KEYOBJECT DIAM NUMBER IN CATALOG BLRD_CATALOG;
-- Quantity of balls
CREATE KEYOBJECT QTY NUMBER IN CATALOG BLRD_CATALOG;
CREATE KEYOBJECT NUM NUMBER IN CATALOG BLRD_CATALOG;   -- Ball number
CREATE KEYOBJECT COLOR CHAR IN CATALOG BLRD_CATALOG;   -- Ball color
CREATE KEYOBJECT BALL{NUM,COLOR} IN CATALOG BLRD_CATALOG;
CREATE KEYOBJECT BALLS{BALL MULTIPLE} IN CATALOG BLRD_CATALOG;
CREATE KEYOBJECT BALLSET{GAME,DIAM,QTY,BALLS} IN BLRD_CATALOG;
```

Then, we create a store `BLRD_STORE` and insert two instances of k-object `BALLSET` as follows.

```
CREATE STORE BLRD_STORE FOR CATALOG BLRD_CATALOG;
INSERT INTO BLRD_STORE INSTANCES
{BALLSET:{GAME:'Carom',DIAM:61.5,QTY:3,
        BALLS:{BALL:{NUM:NULL,COLOR:'red'},
              BALL:{NUM:NULL,COLOR:'white'},
              BALL:{NUM:NULL,COLOR:'white_spot'}}}},
{BALLSET:{GAME:'Pool',DIAM:57.15,QTY:15,
        BALLS:{BALL:{NUM:1, COLOR:'yellow'},
```

```
          BALL:{NUM:2, COLOR:'blue'},
          BALL:{NUM:3, COLOR:'red'},
          BALL:{NUM:4, COLOR:'pink'},
          BALL:{NUM:5, COLOR:'orange'},
          BALL:{NUM:6, COLOR:'green'},
          BALL:{NUM:7, COLOR:'tan'},
          BALL:{NUM:8, COLOR:'black'},
          BALL:{NUM:9, COLOR:'yellow_white'},
          BALL:{NUM:10,COLOR:'blue_white'},
          BALL:{NUM:11,COLOR:'red_white'},
          BALL:{NUM:12,COLOR:'pink_white'},
          BALL:{NUM:13,COLOR:'orange_white'},
          BALL:{NUM:14,COLOR:'green_white'},
          BALL:{NUM:15,COLOR:'tan_white'}}}};
```

Since the Carom balls do not carry numbers, the respective instances of k-object BALL have values of the k-object NUM set to NULL, like in SQL.

MCK: Minimal Containing K-object rule

Consider a simple query with WHERE clause as follows.

```
SELECT *
FROM BLRD_STORE
WHERE COLOR = 'white'
```

It is intuitive relative to the same query without WHERE clause which returns all host instances of the store. The WHERE clause restricts the output host instances to those having a value of k-object COLOR as 'white'. Since only the first host instance satisfies this condition, it will comprise the result as follows.

```
{BALLSET:{GAME:'Carom',DIAM:61.5,QTY:3,
         BALLS:{BALL:{NUM:NULL,COLOR:'red'},
               BALL:{NUM:NULL,COLOR:'white'},
               BALL:{NUM:NULL,COLOR:'white_spot'}}}}
```

The next query looks simple as well.

```
SELECT NUM
FROM BLRD_STORE
WHERE COLOR = 'red'
```

However, it generally may mean two different things as follows.

1) *Get the numbers of all red balls from each* BALLSET.
2) *Get the numbers of all balls from each* BALLSET *that has a red ball.*

The KeySQL semantics of this query corresponds to the first choice according to the following *minimal containing k-object (MCK)* rule.

KeySQL always picks the query scope as the scope of the minimal k-object containing both NUM and COLOR, in our case. Here, the MCK is BALL. Therefore, the values of NUM are taken from all red ball instances contained in each of the host instances in the store. In our example, each of the two host instances contains one red ball, so the resulting store is comprised of the two instances of k-object NUM as follows.

```
{NUM:NULL},
{NUM:3}
```

If we need to answer the query specified by the second interpretation above, it can be done as well. We will return to this example in the section dealing with joins.

To get a better intuition behind the MCK rule, consider rewriting our query as follows.

```
SELECT BALL.NUM
FROM BLRD_STORE
WHERE BALL.COLOR = 'red'
```

The qualified and the original non-qualified query must have the same meaning because NUM and COLOR both have BALL as their parent. And since the BALL in BALL.NUM and in BALL.COLOR should relate to the same instance of BALL, only the red balls are considered for the result.

Partially qualified k-objects

Consider the query:

```
SELECT BALL.COLOR
FROM BLRD_STORE
WHERE BALL.COLOR = 'blue'
```

Like we mentioned at the end of the previous section, the reference to BALL in the SELECT clause and the WHERE clause must mean one and the same instance of BALL, and the answer to the query is:

```
{COLOR:'blue'}
```
Consider now the query:

```
SELECT COLOR
FROM BLRD_STORE
WHERE BALL.COLOR = 'blue'
```

Note that in the SELECT clause, the COLOR is not qualified by its parent BALL but in the WHERE clause it is qualified. In this case, we say that the k-object COLOR is *partially qualified* in the query. Another case of the partial qualification is demonstrated by the following query.

```
SELECT BALL.COLOR
FROM BLRD_STORE
WHERE COLOR = 'blue'
```

Here, in the SELECT clause the COLOR is qualified by its parent BALL, and in the WHERE clause it is not qualified.

The point is that both the second query and the third query are equivalent to the first one. This becomes clear from the observations at the end of the previous section repeated in the beginning of the current section. Summarizing, in KeySQL, *a partial qualification is always equivalent to the full qualification* independently of the query or the store.

To make this point clearer, suppose that some host instance of the BLRD_STORE references two different k-objects: BALL and NOT_BALL, both containing COLOR as an element. So, what is the result of the following query?

```
SELECT COLOR
FROM BLRD_STORE
WHERE NOT_BALL.COLOR = 'red' OR BALL.COLOR = 'blue'
```

Before processing any query, KeySQL server analyzes the query and automatically replaces all partially qualified k-object references with the fully qualified ones.

Since the COLOR in the SELECT clause cannot be qualified by BALL and NOT_BALL at the same time, this query has no answer independently of the WHERE clause content and will produce an error as follows.

> `Error: ambiguous reference to COLOR`

Please note that the next query is valid and will not cause the error because k-object COLOR is not qualified here as opposed to being partially qualified like in the previous query.

```
SELECT COLOR
```

```
FROM BLRD_STORE
WHERE COLOR = 'red' OR COLOR = 'blue'
```

The concept of partial qualification extends to the parent k-objects. Consider the following query with two elements in the select list as follows.

```
SELECT BALL, NOT_BALL
FROM BLRD_STORE
WHERE COLOR = 'blue'
```

Here, both elements of the select list BALL and NOT_BALL contain COLOR as their element. This makes the WHERE clause condition and the whole query ambiguous for the same reason as above: COLOR cannot be qualified by BALL and NOT_BALL at the same time. This too generates the ambiguous reference error.

KeySQL query evaluation algorithm

> The reader can skip this section in the first reading as it represents a deeper dive into KeySQL semantics.

Let us now formally define the MCK concept and then use it to describe the general algorithm of KeySQL query evaluation.

Consider a k-object H that comprises one or more occurrences of each of k-objects a, b, ..., c.

> Within H, a keyobject K is an MCK for the set of k-objects {a, b, ..., c} if K contains each of them and there is no other ("smaller") k-object within K that also contains each of them.

Examples

FIRST_NAME is an MCK for the set of k-objects {FIRST_NAME}.

PERSON{FIRST_NAME,LAST_NAME,ADDRESS} is an MCK for the set {FIRST_NAME, LAST_NAME}.

K-object BALLS{BALL MULTIPLE} is not an MCK for {BALL}, BALL is.

The reply to a query:

```
SELECT S1,S2,…
FROM some_store
WHERE P(x,y,…,z)
```

is constructed as follows.

If none of the host k-objects in the store contains all k-objects referenced in a query, namely all k-objects from the set:

{S1, S2, ..., x, y, ..., z},

an error is generated as follows:

```
Error: query doesn't match any host
```

Step 0. If any k-object referenced in the query is partially qualified, rewrite the query to make it fully qualified. If this cannot be done unambiguously return an error.

After this step, the query processing can be performed separately for each element of the SELECT list and the reply to the original query is formed as the composition of the respective separate replies.

For each of the k-objects in the original SELECT list, the reply to a query:

```
SELECT S
FROM some_store
WHERE P(x,y,…,z);
```

is constructed as follows.

Step 1. For each host k-object H containing all k-objects from the set {S,x, y, ..., z}, find all MCKs for the set of k-objects {S, x, y, ..., z}.

Step 2. In each instance of each host k-object H of the store, find all instances of MCKs of Step 1 for which the predicate P(x, y, ..., z) is true.

Step 3. For each MCK of Step 1 that has a single occurrence of k-object S, select the instance of S from each MCK instance found at Step 2. All these instances of S fall into the query reply.

Step 4. For each MCK of Step 1 that has multiple occurrences of k-object S, for each instance M of the MCK found at Step 2, select all instances of S from M for each of which the predicate P(x, y, ..., z) is true on M. All these instances of S fall into the query reply.

Special cases

Case 1

The query:

```
SELECT *
FROM some_store
```

```
WHERE P(x,y,…,z);
```

Is equivalent to the UNION ALL of the queries:

```
SELECT H
FROM some_store
WHERE P(x,y,…,z);
```

where H is a host of the store and, of course, the MCK.

Case 2

The queries:

```
SELECT _IID
FROM some_store
WHERE P(x,y,…,z);
```

and

```
SELECT _VERSION
FROM some_store
WHERE P(x,y,…,z);
```

are processed the same way as the query:

```
SELECT *
FROM some_store
WHERE P(x,y,…,z);
```

But they return just _IID or _VERSION respectively, for each host instance.

The presence of any number of occurrences of _IID or _VERSION in the WHERE clause does not affect the selection of the MCKs for any query. They are only accounted for in verifying the predicate P as the additional filters.

Instance markers

Consider a query:

```
SELECT S
FROM some_store
WHERE P(x,y,…,z);
```

In SQL it means: Select the value of column S from each row of the table where the values in columns x, y, …, z of the row satisfy the predicate P of the WHERE clause. If a column is referenced in the WHERE clause more than once, each reference relates to one and the same value located in one and the same cell of the column. And because of that a single variable (column name) is always sufficient for any number of the column references in the WHERE clause.

Unlike the rows in SQL tables, k-objects can have multiple instances within one host instance. If a k-object name is referenced more than once, then it must relate to the same k-object instance, analogously to SQL. But what if we need to express a query that must relate to possibly different instances of a k-object within the same host instance of the store? In this case we must use more than one variable in the query for the instances of the k-object. In KeySQL, the names for the respective variables can be syntactically constructed using the *instance markers*.

An *instance marker* is an identifier in square brackets located after the name of a k-object that is contained as MULTIPLE in the definition of another "bigger" k-object.

Consider the BLRD_CATALOG example and the following query addressed to the BLRD_STORE.

```
SELECT GAME
FROM BLRD_STORE
WHERE COLOR = 'red' AND NUM = 15
```

This query is of course equivalent to the following one.

```
SELECT GAME
FROM BLRD_STORE
WHERE BALL.COLOR = 'red' AND BALL.NUM = 15
```

And since one and the same variable BALL is used for COLOR and NUM, the meaning of the query is: *find all games that have red balls with number 15.* The reply to this query is empty because there is no game where a red ball carries number 15.

The same query can be expressed using the instance markers as follows.

```
SELECT GAME
FROM BLRD_STORE
WHERE BALL[b].COLOR = 'red' AND BALL[b].NUM = 15
```

The two identical instance markers [b] mean that the values of k-objects COLOR and NUM must both occur in the same instance of BALL just like in the previous query. If the markers are different, it means that the values of COLOR and NUM may also come from different instances of BALL.

This query can be expressed as follows.

```
SELECT GAME
FROM BLRD_STORE
WHERE BALL[b1].COLOR = 'red' AND BALL[b2].NUM = 15
```

The reply now is as follows.

```
{GAME:'Pool'}
```

The only objective of appending *instance* markers to the qualifying k-object names is to specify that the WHERE clause references either one and the same or potentially different instances of k-object BALL. In lieu of [b] or [b1] and [b2], we can use any identifier provided it is the same or, respectively, different for the occurrences of BALL. For example, the following query is equivalent to the previous one.

```
SELECT GAME
FROM BLRD_STORE
WHERE BALL[X].COLOR = 'red' AND BALL[Y].NUM = 15
```

Along the same line, consider the following query.

```
SELECT GAME
FROM BLRD_STORE
WHERE COLOR = 'red' AND COLOR = 'white'
```

In SQL, the above WHERE clause condition will always evaluate to FALSE since there is only one element in any table cell, and a color cannot be red and not red at the same time. The same happens in KeySQL, and therefore the result of the query is empty.

However, the following modification of the query will return a non-empty reply.

```
SELECT GAME
FROM BLRD_STORE
WHERE BALL[b1].COLOR = 'red' AND BALL[b2].COLOR = 'white'
```

And the answer is as follows.

```
{GAME:'Carom'}
```

There also can be a query whereby we are looking for the games that have red balls and balls with number 15 but those must be different BALL instances. This semantics is not currently implemented with instance markers. However, the answer to this query can be obtained using the set operator MINUS as follows.

```
SELECT GAME
FROM BLRD_STORE
WHERE BALL[b1].COLOR = 'red' AND BALL[b2].NUM = 15
MINUS
SELECT GAME
FROM BLRD_STORE
WHERE COLOR = 'red' AND NUM = 15
```

Query evaluation examples

To demonstrate the KeySQL query evaluation algorithm specified above, let us consider a couple of query examples as follows. We recommend that the reader follows the algorithm when reviewing explanations of the examples.

Example 1

```
SELECT BALL
FROM BLRD_STORE
WHERE COLOR = 'red' AND GAME ='Carom'
```

BLRD_STORE has a single host k-object BALLSET, which is the MCK for {BALL, COLOR, GAME}. BALLSET has multiple occurrences of BALL.

Therefore, the query is:

> From each instance of Ballset for which the WHERE clause is true, select all instances of Ball for which the WHERE clause is true on this instance of Ballset.

It produces the following resulting store.

```
{BALL: {NUM:NULL, COLOR:'red'}}
```

Example 2

```
SELECT NUM
FROM BLRD_STORE
WHERE NUM + DIAM = 58.15;
```

BLRD_STORE has a single host k-object BALLSET, which is the MCK for {NUM, DIAM}. BALLSET has multiple occurrences of Num (each of which belongs to an occurrence of BALL, but DIAM does not belong to BALL).

Therefore, the query is:

> From each instance of Ballset for which the WHERE clause is true, select all instances of Num for which the WHERE clause is true on this instance of Ballset.

It produces the following resulting store.

```
{NUM: 1}
```

1.14   Using operator AS for flattening

The Billiard data store example allows to demonstrate how the AS operator can be used to flatten k-object instances containing multiples.

Imagine that we want to have a flat, table-like representation of the BALLSET instances to be able to present them as a spreadsheet or as a report using some BI tool. For example, we want each row of the "table" to show the following "columns": game, ball number, ball color, and the ball diameter.

To do that we create a new k-object FLAT_BALLSET as follows.

```
CREATE KEYOBJECT FLAT_BALLSET{GAME,NUM,COLOR,DIAM} IN CATALOG
BLRD_CATALOG;
```

Now to get our "table" we execute the following query with AS operator. The WHERE clause filter is arbitrary and is used just to shorten the output, since without it the result will contain 18 instances.

```
SELECT {GAME,NUM,COLOR,DIAM} AS FLAT_BALLSET
FROM BLRD_STORE
WHERE (NUM IS NULL) AND (COLOR='red' OR COLOR='white') OR NUM < 3
ORDER BY GAME, NUM
```

The query reply is shown below.

```
{FLAT_BALLSET: {GAME:'Carom', DIAM:61.5, NUM:NULL, COLOR:'red'}},
{FLAT_BALLSET: {GAME:'Carom', DIAM:61.5, NUM:NULL, COLOR:'white'}},
{FLAT_BALLSET: {GAME:'Pool', DIAM:57.15, NUM:1, COLOR:'yellow'}},
{FLAT_BALLSET: {GAME:'Pool', DIAM:57.15, NUM:2, COLOR:'blue'}}
```

Using the SET OUTPUT_FORMAT statement, the reply data set above can be automatically returned by KeySQL server as a JSON file, which then can be fed to a BI tool or machine learning software, for example.

The JSON text of the query output looks as follows.

```
[{"FLAT_BALLSET":
{"GAME":"Carom","DIAM":61.5,"NUM":null,"COLOR":"red"}},
{"FLAT_BALLSET":
{"GAME":"Carom","DIAM":61.5,"NUM":null,"COLOR":"white"}},
{"FLAT_BALLSET":
{"GAME":"Pool","DIAM":57.15,"NUM":1,"COLOR":"yellow"}},
{"FLAT_BALLSET":
{"GAME":"Pool","DIAM":57.15,"NUM":2,"COLOR":"blue"}}]
```

This data set can also be exported to CSV or another spreadsheet compatible format.

1.15   User-defined ad hoc k-objects

The TO and AS operators were introduced as means of renaming and restructuring query results by using catalog k-objects, particularly in lieu of standard ad hoc k-objects like RESULT. The price of employing the operators is the effort of defining the needed k-objects in a catalog so they can be used as the target data structures for the TO or AS operators.

This effort is unavoidable when we want to save the query results using the INSERT SELECT statement, since it is impossible to insert an instance of a k-object into a store if it does not exist in the respective catalog. However, in the use cases where there is no intent of saving the query results, the need to predefine the target k-objects specifically for the TO or AS operators becomes a pure overhead.

In this section we introduce a concept of *user-defined ad hoc k-objects* which eliminate this overhead when there is no intent of persisting the query results.

Consider the familiar query example as follows.

```
SELECT { LAST_NAME } TO LAST_NAMES
FROM MY_STORE
```

For this query to run without errors we had to first add k-object LAST_NAMES to the catalog MY_CATALOG since the store MY_STORE is based on this catalog.

Instead of a catalog k-object, like LAST_NAMES, we can employ a user-defined ad hoc k-object and KeySQL Server will automatically infer its structure. To tell the server we are using an ad-hoc k-object, its name must be prefixed with the dollar sign. The name itself does not matter and can even coincide with the standard ad hoc k-objects like RESULT.

For example, we can choose to use an ad hoc keyobject $LNS as follows.

```
SELECT { LAST_NAME } TO $LNS
FROM MY_STORE
```

And the result as follows.

```
{LNS: {LAST_NAME:'Johnson'}},
{LNS: {LAST_NAME:'Doe'}},
{LNS: {LAST_NAME:'Smith', LAST_NAME:'Brown'}}
```

We can choose the ad hoc $LAST_NAMES as well.

```
SELECT { LAST_NAME } TO $LAST_NAMES
FROM MY_STORE
```

And we get the familiar result.

```
{LAST_NAMES: {LAST_NAME:'Johnson'}},
{LAST_NAMES: {LAST_NAME:'Doe'}},
{LAST_NAMES: {LAST_NAME:'Smith', LAST_NAME:'Brown'}}
```

If the k-object LAST_NAMES is not identically defined in MY_CATALOG this result will be *uninsertable*, which means the query will cause an error in the INSERT SELECT statement. To make it insertable we need to create the k-object in the catalog.

Though a k-object with the same name as the ad hoc may already exist in the catalog, KeySQL Server will always use the ad hoc one. So, if the structure of the catalog k-object and the ad hoc one with the same name are different, then, again, the result will be uninsertable.

Several user-defined ad hoc k-objects can be used in a query. For example, consider another familiar query as follows.

```
SELECT {{LAST_NAME,FIRST_NAME} AS $PERSON} AS $PEOPLE
FROM MY_STORE
```

The result is of course as follows.

```
{PEOPLE: {PERSON:{LAST_NAME:'Doe', FIRST_NAME: 'Jane'}}},
{PEOPLE: {PERSON:{LAST_NAME:'Smith', FIRST_NAME:'John'}},
         {PERSON:{LAST_NAME:'Brown',FIRST_NAME:'Jane'}}}
```

1.16   SELECT without FROM

In KeySQL, SELECT statement without FROM clause accepts only constant expressions and produces a single k-object instance, which can be either elementary, like a number, character string, or a date, or an instance of a compound k-object.

Let us start with the following query examples that produce instances of elementary k-objects.

```
SELECT NULL
{RESULT: NULL}

SELECT 1
{RESULT: 1}

SELECT EXTRACT(Year FROM CURRENT_DATE())
{RESULT : 2021}

SELECT 'sin(1)='||sin(1)
{RESULT: 'sin(1)=0.841470984807897'}

SELECT LN(10) + 1 AS $Z    -- Note user-defined ad hoc k-object $Z
{Z: 3.30258509299405}
```

Here are examples of SELECT statements producing instances of compound k-objects which employ user-defined ad hoc k-objects.

```
SELECT 1 TO $K,'abc' AS $A
{RESULT: {K: 1, A: 'abc'}}

SELECT {'a' AS $A1,'abc' AS $A2} AS $AA
{AA: {A1: 'a', A2 : 'abc'}}

SELECT {{'a' AS $A1,'abc' AS $A2} AS $AA} TO $AA2
{AA2: {AA: {A1: 'a', A2: 'abc' }}}

SELECT {{1 as $A} as $B1, {1 as $a} as $b2} as $c
{C: {B1: {A: 1}, B2: {A: 1}}}
```

> Note: Constant expressions are not allowed in SELECT statements with FROM clause

For example, the following statement will cause an error.

```
SELECT 1
FROM MY_STORE
```

1.17   Set operators

In SQL, set operators (operations) can only be used for tables having the same number of columns of the compatible data types (union-compatible tables). In KeySQL, set operations can be performed on arbitrary stores independently of the data structure of their hosts. The operand stores are formed as the results of SELECT statements. The operand stores must be defined for the same catalog.

Set operations are possible only for SELECT statements that produce either instances of catalog k-objects, or instances of user-defined ad hoc k-objects. They are not valid for standard default k-objects like RESULT.

The set operators are as follows.

UNION ALL

UNION

INTERSECT

MINUS or EXCEPT

UNION ALL

The UNION ALL of two stores formed as the results of the respective SELECT statements is a store that includes all host instances from the first store and all host instances from the second store. This operation is commutative – the result does not depend on the order of the operands.

Example

```
SELECT AGE FROM MY_STORE LIMIT 100
UNION ALL
SELECT AGE FROM MY_STORE
UNION ALL
SELECT FIRST_NAME AS FIRST_NAME FROM MY_STORE
```

The result is as follows.

```
{AGE:20}, {AGE:20}, {FIRST_NAME:'Jane'}, {FIRST_NAME:'John'},
{FIRST_NAME:'Jane'}
```

UNION

The UNION of two stores formed as the results of the respective SELECT statements is a store that includes all host instances from the first store and all host instances from the second store. All duplicates are eliminated. This operation is commutative – the result does not depend on the order of the operands.

Example

```
SELECT AGE FROM MY_STORE LIMIT 100
UNION
SELECT AGE FROM MY_STORE
UNION
SELECT FIRST_NAME AS FIRST_NAME FROM MY_STORE
```

The result is as follows.

```
{AGE:20}, {FIRST_NAME:'Jane'}, {FIRST_NAME:'John'}
```

INTERSECT

The INTERSECT of two stores formed as the results of the respective SELECT statements is a store that includes all common host instances from the first store and the second store. All duplicates are eliminated. This operation is commutative – the result does not depend on the order of the operands.

Example

```
SELECT FIRST_NAME AS FIRST_NAME FROM MY_STORE(PEOPLE)
INTERSECT
SELECT FIRST_NAME AS FIRST_NAME FROM MY_STORE(PERSON)
```

The result is as follows.

```
{FIRST_NAME:'Jane'}
```

MINUS or EXCEPT

The MINUS, also called EXCEPT, of two stores formed as the results of the respective SELECT statements is a store that includes all host instances from the first store that are not present in the second store. Al duplicates are eliminated. Like the minus operation in arithmetic, this operation is not commutative. The result depends on the order of the operands.

Example

```
SELECT FIRST_NAME AS FIRST_NAME FROM MY_STORE(PEOPLE)
MINUS
SELECT FIRST_NAME AS FIRST_NAME FROM MY_STORE(PERSON)
```

The result is as follows.

```
{FIRST_NAME:'John'}
```

1.18   JOIN clause

In SQL, JOIN clause is used to combine columns from one or more tables according to the join condition, and there are several types of joins. In KeySQL, joins allow to combine k-object instances from one or more stores based on the same catalog, and currently there are two types of joins: JOIN and LEFT JOIN.

The syntax is very close to the one of SQL, but the join semantics is generalized to accommodate the arbitrarily complex data objects of KeySQL. Naturally, KeySQL joins can produce data layouts that cannot be expressed within SQL.

Suppliers-Parts data stores

To introduce the join functionality, we use a simple data set of Suppliers-Parts stores. It is compiled from the classic Supplier-Parts relational database example employed by Chris Date in his book "*Introduction to Database Systems*" (*6th edition, Addison -Wesley, 1995*). The goal of using these "flat" data stores is for the reader to better understand the specifics of KeySQL joins as compared to the relational ones. As we shall see, though the base data stores are flat, the JOIN results, just like the GROUP BY results, may have a richer structure.

First, we create a SUPP_PART catalog with the required k-objects as follows.

```
CREATE CATALOG SUPP_PART;
CREATE KEYOBJECT S_ID CHAR IN CATALOG SUPP_PART;     -- Supplier ID
CREATE KEYOBJECT P_ID CHAR IN CATALOG SUPP_PART;     -- Part ID
CREATE KEYOBJECT S_NAME CHAR IN CATALOG SUPP_PART; -- Supplier name
CREATE KEYOBJECT P_NAME CHAR IN CATALOG SUPP_PART; -- Part name
CREATE KEYOBJECT COLOR CHAR IN CATALOG SUPP_PART;    -- Part color
CREATE KEYOBJECT CITY CHAR IN CATALOG SUPP_PART;
CREATE KEYOBJECT STATUS NUMBER IN CATALOG SUPP_PART;
CREATE KEYOBJECT WEIGHT NUMBER IN CATALOG SUPP_PART;
```

```
-- Supplier
CREATE KEYOBJECT S{S_ID,S_NAME,STATUS,CITY} IN CATALOG SUPP_PART;
-- Part
CREATE KEYOBJECT P{P_ID,P_NAME,COLOR,WEIGHT,CITY} IN CATALOG
SUPP_PART;
```

We also create k-object S_P representing a pair <Supplier, Part> that will be used in our examples.

```
CREATE KEYOBJECT S_P{S,P} IN CATALOG SUPP_PART;
```

Let us now create two data stores matching the respective tables of Chris Date and populate them with k-object instances that correspond to the table rows as follows.

```
CREATE STORE SUPPLIERS FOR CATALOG SUPP_PART;
INSERT INTO SUPPLIERS INSTANCES
{S: {S_ID:'S1', S_NAME:'Smith', STATUS:20, CITY:'London'}},
{S: {S_ID:'S2', S_NAME:'Jones', STATUS:10, CITY:'Paris'}},
{S: {S_ID:'S3', S_NAME:'Blake', STATUS:30, CITY:'Paris'}},
{S: {S_ID:'S4', S_NAME:'Clark', STATUS:20, CITY:'London'}},
{S: {S_ID:'S5', S_NAME:'Clark', STATUS:20, CITY:'Athens'}};

CREATE STORE PARTS FOR CATALOG SUPP_PART;
INSERT INTO PARTS INSTANCES
{P: {P_ID:'P1', P_NAME:'Nut', COLOR:'Red', WEIGHT:12,
CITY:'London'}},
{P: {P_ID:'P2', P_NAME:'Bolt', COLOR:'Green', WEIGHT:17,
CITY:'Paris'}},
{P: {P_ID:'P3', P_NAME:'Screw', COLOR:'Blue', WEIGHT:17,
CITY:'Rome'}},
{P: {P_ID:'P4', P_NAME:'Screw', COLOR:'Red', WEIGHT:14,
CITY:'London'}},
{P: {P_ID:'P5', P_NAME:'Cam', COLOR:'Blue', WEIGHT:12,
CITY:'Paris'}},
{P: {P_ID:'P6', P_NAME:'Cog', COLOR:'Red', WEIGHT:19,
CITY:'London'}};
```

SQL-like joins

Let us start with the following query example where we use the flat k-object S_P to present the result as the pair <Supplier, Part> almost like it is happening in SQL. The difference consists in referencing the whole entities, Supplier and Part, corresponding to SQL tables rather than referencing all the columns of the respective tables.

```
SELECT {S,P} AS S_P
FROM SUPPLIERS SPL JOIN PARTS PRT ON SPL.CITY = PRT.CITY
WHERE COLOR = 'Red' AND P_NAME <> 'Screw';
```

It produces the following four instances as the result:

```
{S_P: {S: {S_ID:'S1', S_NAME:'Smith', CITY:'London', STATUS:20},
       P: {P_ID:'P1', P_NAME:'Nut', COLOR:'Red', CITY:'London',
WEIGHT:12}}},
{S_P: {S: {S_ID:'S1', S_NAME:'Smith', CITY:'London', STATUS:20},
       P: {P_ID:'P6', P_NAME:'Cog', COLOR:'Red', CITY:'London',
WEIGHT:19}}},
{S_P: {S: {S_ID:'S4', S_NAME:'Clark', CITY:'London', STATUS:20},
       P: {P_ID:'P1', P_NAME:'Nut', COLOR:'Red', CITY:'London',
WEIGHT:12}}},
{S_P: {S: {S_ID:'S4', S_NAME:'Clark', CITY:'London', STATUS:20},
       P: {P_ID:'P6', P_NAME:'Cog', COLOR:'Red', CITY:'London',
WEIGHT:19}}}
```

The analog of this query in SQL can look as follows.

```
SELECT SPL.*, PRT.*
FROM SUPPLIERS SPL JOIN PARTS PRT ON SPL.CITY = PRT.CITY
WHERE COLOR = 'Red' AND P_NAME <> 'Screw'
```

Just like in SQL, if we change the order of the join arguments, the respective query below produces the same result.

```
SELECT {S,P} AS S_P
FROM PARTS PRT JOIN SUPPLIERS SPL ON SPL.CITY = PRT.CITY
WHERE COLOR = 'Red' AND P_NAME <> 'Screw'
```

We can also get the same output with a query that uses GROUP BY clause as follows.

```
SELECT S,P
FROM SUPPLIERS SPL JOIN PARTS PRT ON SPL.CITY = PRT.CITY
WHERE COLOR = 'Red' AND P_NAME <> 'Screw'
GROUP BY S,P
```

In this case the output is presented using the *ad hoc* k-object RESULT as follows.

```
{RESULT: {S: {S_ID:'S1', S_NAME:'Smith', CITY:'London', STATUS:20},
          P: {P_ID:'P1', P_NAME:'Nut', COLOR:'Red', CITY:'London',
WEIGHT:12}}},
{RESULT: {S: {S_ID:'S1', S_NAME:'Smith', CITY:'London', STATUS:20},
```

```
          P: {P_ID:'P6', P_NAME:'Cog', COLOR:'Red', CITY:'London',
WEIGHT:19}}},
{RESULT: {S: {S_ID:'S4', S_NAME:'Clark', CITY:'London', STATUS:20},
          P: {P_ID:'P1', P_NAME:'Nut', COLOR:'Red', CITY:'London',
WEIGHT:12}}},
{RESULT: {S: {S_ID:'S4', S_NAME:'Clark', CITY:'London', STATUS:20},
          P: {P_ID:'P6', P_NAME:'Cog', COLOR:'Red',
CITY:'London',WEIGHT:19}}}
```

And again, the order of the join arguments does not matter.

When join order matters

What if we remove GROUP BY clause from the previous query as follows?

```
SELECT S,P
FROM SUPPLIERS SPL JOIN PARTS PRT ON SPL.CITY = PRT.CITY
WHERE COLOR = 'Red' AND P_NAME <> 'Screw'
```

The query produces the following two instances as the result.

```
{RESULT: {S: {S_ID:'S1', S_NAME:'Smith', CITY:'London', STATUS:20},
   #P: {P: {P_ID:'P1', P_NAME:'Nut', COLOR:'Red', CITY:'London',
WEIGHT:12},
        P: {P_ID:'P6', P_NAME:'Cog', COLOR:'Red', CITY:'London',
WEIGHT:19}}}},

{RESULT: {S: {S_ID:'S4', S_NAME:'Clark', CITY:'London', STATUS:20},
   #P: {P: {P_ID:'P1', P_NAME:'Nut', COLOR:'Red', CITY:'London',
WEIGHT:12},
        P: {P_ID:'P6', P_NAME:'Cog', COLOR:'Red', CITY:'London',
WEIGHT:19}}}}
```

In this layout, parts are automatically grouped by suppliers rather than being repeated along with the suppliers. The output becomes shorter and arguably more readable since we do not need to eye-ball the results to distinguish between the records related to different suppliers.

Let us alter the order of the arguments in the JOIN clause as follows.

```
SELECT S,P
FROM PARTS PRT JOIN SUPPLIERS SPL ON SPL.CITY = PRT.CITY
WHERE COLOR = 'Red' AND P_NAME <> 'Screw'
```

Now, suppliers are automatically grouped by parts as follows.

```
{RESULT: {#S: {S: {S_ID:'S1', S_NAME:'Smith', CITY:'London',
STATUS:20},
              S:{S_ID:'S4', S_NAME:'Clark', CITY:'London',
STATUS:20}},
         P:{ P_ID:'P1', P_NAME:'Nut', COLOR:'Red', CITY:'London',
WEIGHT:12}}},

{RESULT: {#S: {S: {S_ID:'S1', S_NAME:'Smith', CITY:'London',
STATUS:20},
              S: {S_ID:'S4', S_NAME:'Clark', CITY:'London',
STATUS:20}},
         P: {P_ID:'P6', P_NAME:'Cog', COLOR:'Red', CITY:'London',
WEIGHT:19}}}
```

To first list the part and then the suppliers within the instances of RESULT, we simply change their order in the SELECT list as follows.

```
SELECT P,S
FROM PARTS PRT JOIN SUPPLIERS SPL ON SPL.CITY = PRT.CITY
WHERE COLOR = 'Red' AND P_NAME <> 'Screw'
```

This makes the trick, and we get the following output:

```
{RESULT:{P:{P_ID:'P1',P_NAME:'Nut',COLOR:'Red',CITY:'London',
        WEIGHT:12},
        #S:{S:{S_ID:'S1',S_NAME:'Smith',CITY:'London',STATUS:20},
            S:{S_ID:'S4',S_NAME:'Clark',CITY:'London',STATUS:20}}}},

{RESULT:{P:{P_ID:'P6',P_NAME:'Cog',COLOR:'Red',CITY:'London',
        WEIGHT:19},
        #S:{S:{S_ID:'S1',S_NAME:'Smith',CITY:'London',STATUS:20},
            S:{S_ID:'S4',S_NAME:'Clark',CITY:'London',STATUS:20}}}}
```

Summarizing, to get the longest and symmetrical join output we need to either use GROUP BY or AS operator with a symmetric k-object, like <Supplier, Part>. This produces the "normalized" list of pairs of all suppliers and parts from each city.

The other two options are:

1) The list of parts grouped by suppliers.

2) The list of suppliers grouped by parts.

The choice between these two options is controlled by the order of the join operands.

LEFT JOIN

In SQL, LEFT JOIN is asymmetric, and the same is true for KeySQL. Let us consider an example of a left join query, which at the same time represents a category of joins called *non-equi* joins. This means the join condition is other than equality. The query is as follows.

```
SELECT {S,P} AS S_P
FROM SUPPLIERS SPL LEFT JOIN PARTS PRT
    ON SPL.STATUS >= 2.5*PRT.WEIGHT
WHERE SPL.CITY <> 'London'
```

It produces a resulting store consisting of four host instances as follows.

```
{S_P: {S: {S_ID:'S2', S_NAME:'Jones', CITY:'Paris', STATUS:10},
       P: NULL}},
{S_P: {S: {S_ID:'S3', S_NAME:'Blake', CITY:'Paris', STATUS:30},
       P: {P_ID:'P1', P_NAME:'Nut', COLOR:'Red', CITY:'London',
WEIGHT:12}}},
{S_P: {S: {S_ID:'S3', S_NAME:'Blake', CITY:'Paris', STATUS:30},
       P:{ P_ID:'P5', P_NAME:'Cam', COLOR:'Blue', CITY:'Paris',
WEIGHT:12}}},
{S_P: {S: {S_ID:'S5', S_NAME:'Clark', CITY:'Athens', STATUS:20},
       P: NULL}}}
```

Please note that the first and the fourth instance shows the instance of composite k-object P having the value of NULL. This means that all instances of all k-objects it hosts are null. This is different from SQL where we cannot designate an entity as null but rather must display the nulls in all columns of the row.

Self-joins

In SQL, self-join means joining tables to itself. In KeySQL, self-join means joining stores to itself. But, unlike SQL tables, stores can contain host instances of different k-objects, that is the records of different structures. For example, stores can contain instances of suppliers along with instances of parts. Let us create a store like that and call it SUPP_PART_STORE as follows.

```
CREATE STORE SUPP_PART_STORE FOR CATALOG SUPP_PART
```

Then we insert into this store all instances from the SUPPLIERS and PARTS stores.

```
INSERT INTO SUPP_PART_STORE INSTANCES
{S: {S_ID:'S1', S_NAME:'Smith', STATUS:20, CITY: 'London'}},
```

```
{S: {S_ID:'S2', S_NAME:'Jones', STATUS:10, CITY: 'Paris'}},
{S: {S_ID:'S3', S_NAME:'Blake', STATUS:30, CITY: 'Paris'}},
{S: {S_ID:'S4', S_NAME:'Clark', STATUS:20, CITY: 'London'}},
{S: {S_ID:'S5', S_NAME:'Clark', STATUS:20, CITY: 'Athens'}}
{P: {P_ID:'P1', P_NAME:'Nut', COLOR:'Red', WEIGHT: 12,
CITY: 'London'}},
{P: {P_ID:'P2', P_NAME:'Bolt', COLOR:'Green', WEIGHT: 17,
CITY:'Paris'}},
{P: {P_ID:'P3', P_NAME:'Screw', COLOR:'Blue', WEIGHT: 17,
CITY:'Rome'}},
{P: {P_ID:'P4', P_NAME:'Screw', COLOR:'Red', WEIGHT: 14,
CITY:'London'}},
{P: {P_ID:'P5', P_NAME:'Cam', COLOR:'Blue', WEIGHT: 12,
CITY:'Paris'}},
{P: {P_ID:'P6', P_NAME:'Cog', COLOR:'Red', WEIGHT: 19,
CITY:'London'}};
```

Let us now modify our very first join query for the case of self-join as follows.

```
SELECT {SPL.S,PRT.P} AS S_P
FROM SUPP_PART_STORE(S) SPL JOIN SUPP_PART_STORE(P) PRT
     ON SPL.CITY = PRT.CITY
WHERE PRT.COLOR = 'Red' AND PRT.P_NAME <> 'Screw'
```

This query uses scope to distinguish suppliers and parts in SUPP_PART_STORE and produces the same result as the original query addressed to two stores, as follows.

```
{S_P: {S: {S_ID:'S1', S_NAME:'Smith', CITY:'London', STATUS:20},
       P: {P_ID:'P1', P_NAME:'Nut', COLOR:'Red', CITY:'London',
WEIGHT:12}}},
{S_P: {S: {S_ID:'S1', S_NAME:'Smith', CITY:'London', STATUS:20},
       P: {P_ID:'P6', P_NAME:'Cog', COLOR:'Red', CITY:'London',
WEIGHT:19}}},
{S_P: {S: {S_ID:'S4', S_NAME:'Clark', CITY:'London', STATUS:20},
       P: {P_ID:'P1', P_NAME:'Nut', COLOR:'Red', CITY:'London',
WEIGHT:12}}},
{S_P: {S: {S_ID:'S4', S_NAME:'Clark', CITY:'London', STATUS:20},
       P: {P_ID:'P6', P_NAME:'Cog', COLOR:'Red', CITY:'London',
WEIGHT:19}}}
```

Self-joins do not necessarily need to use scope. For this query we get the same result by rewriting it without the scope as follows.

```
SELECT {SPL.S,PRT.P} AS S_P
FROM SUPP_PART_STORE SPL JOIN SUPP_PART_STORE PRT
     ON SPL.CITY = PRT.CITY
WHERE PRT.COLOR = 'Red' AND PRT.P_NAME <> 'Screw';
```

Consider now a query that was mentioned when discussing the MCK rule with the Billiard data store example. The query is: *get the numbers of all balls from each* BALLSET *that has a red ball.* It can be answered using a self-join as follows.

```
SELECT X.NUM
FROM BLRD_STORE X JOIN BLRD_STORE Y ON X.GAME = Y.GAME
WHERE Y.COLOR = 'red';
```

We get the following result.

```
{#NUM: {NUM:NULL, NUM:NULL, NUM:NULL}},
{#NUM: {NUM:1, NUM:2, NUM:3, NUM:4, NUM:5, NUM:6, NUM:7, NUM:8,
        NUM:9, NUM:10, NUM:11, NUM:12, NUM:13, NUM:14, NUM:15}}
```

Consider a modification of the query which uses the same self-join but lists the ball numbers without the references to their host instances in the store. The DISTINCT clause eliminates the duplicate nulls and the ORDER BY clause orders the numbers in the output. We use the LIMIT clause to shorten the reply to 4 instances.

```
SELECT DISTINCT X.NUM
FROM BLRD_STORE X JOIN BLRD_STORE Y ON X.GAME = Y.GAME
WHERE Y.COLOR = 'red'
ORDER BY X.NUM
LIMIT 4
```

The result is as follows.

```
{NUM: NULL},
{NUM: 1},
{NUM: 2},
{NUM: 3}
```

Security

2

2.1   Overview

Platform and network security

The platform for KeySQL server includes the hardware, the network clients connecting to the servers, and the backend database storage.

KeySQL server client networking supports both secure (TLS) and non-secure connections.

Principals and securables

Principals are the individuals or the roles which are granted access to KeySQL server objects. The *securables* are the server and its objects to which access can be granted. Unless explicitly allowed by a grant, the access is denied.

> Roles will be available in the future versions of KeySQL.

KeySQL server has the multi-level scope hierarchy. The topmost scope is the SERVER (or SYSTEM) within which SCHEMA objects reside. CATALOG objects are contained within the SCHEMA objects. STORE objects are accessed within the SCHEMA scope as well. However, a CATALOG object is required when creating a STORE.

```
SERVER (SYSTEM) scope
|--- SCHEMA
     |--- CATALOG
          |--- STORE
```

The following securables are contained in the SYSTEM scope:

- USER
- PASSWORD POLICY
- ROLE (not yet implemented)
- LDAP LINK (not yet implemented)
- SCHEMA

The following securables are managed within the SCHEMA scope:

- CATALOG
- STORE

See GRANT and REVOKE statements for more details and examples.

2.2   User management

Each user has a login name and a password required for the authentication process.

> The user login names mapping against LDAP (i.e., Active Directory) will be available in the future versions of KeySQL.

The following two users are initially created and are intended for internal use only:

Login name	Initial password	Description
keysqladmin	keysqladmin	Administrator having full server scope rights
keysqluser	Keysqldemo	User having full access to the PUBLIC and KEYSQLUSER schemas

The statements to manage users and other server objects will be described below in the section "User management statements". The passwords above must be changed for normal operation.

2.3   Permissions

Access checking and propagation

Access checking is bottom-up: the securable record, then its scope, then the scope of scope (if applicable). Permissions should be checked bottom-up: from STORE level up to the SCHEMA level.

Access propagation is top-down: the scope, then all underlying scopes; and then the securables. This may require creating explicit permissions for the securables.

2.4 Ownership

Every KeySQL object has an owner. By default, the owner is the user that has created the object. Some CREATE statements allow to specify the owner's name explicitly.

The owner has all permissions on the owned object. For example, a schema owner can modify and drop the schema itself, is able to create new catalogs and stores within the schema, and to transfer the ownership to another user.

KeySQL language reference   3

The notation used for a syntax description is Backus-Naur Form (BNF) including several human-readable extensions. The key points are as follows.

- The language words are defined in plain text, i.e., WORD
- The language symbols are defined using quoted strings, i.e., "{", "]"
- The apostrophe symbol is defined as is '
- The quote symbol is defined with apostrophes '"'
- The meta symbol ::= is to be interpreted as "is defined as"
- The pipe symbol | is used to separate alternative definitions and is interpreted as "or"
- The angle brackets < > are delimiters enclosing a class name of basic symbols
- The parentheses ( ) are for grouping
- Square brackets [ ] indicate optional clauses
- M..N means any value from the inclusive (closed) interval between N and M

Basic language elements:

```
<letter> ::= "A".."Z" | "a".."z"
<simple identifier> ::= <letter> | "_" [ <simple identifier> |
<digit> ]
<quoted identifier> ::=
    '"' <any symbol> | <escaped quote> [ <quoted identifier> ] '"'
<escaped quote> ::= '\"'
<identifier> ::= <simple identifier> | <quoted identifier>
```

Commonly used language elements:

```
<comma_separated_option_list> ::=
    <option> [ "," <comma_separated_option_list> ]
```

Where any <option> has a custom definition specific for statement.

External language elements:

```
<ISO_8601_timestamp> ::= ' YYYY-MM-DDThh:mm:ss.sss '
<json_document> ::= any valid JSON document
```

For <ISO_8601_timestamp> see ISO 8601 standard for more details.

For JSON (JavaScript Object Notation) see https://tools.ietf.org/html/rfc8259.

3.1   Separators

Usually, KeySQL clients accept several statements at once (a batch) or a whole KSQL source file. Statement separator in the batch is the semicolon ; symbol.

> A batch separator will be available in the future versions of KeySQL.

3.2   Comments

Single line comments start with --. Any text between -- and the end of the line will be ignored.

Multiline comments start with /* and end with */. Any text between /* and */ will be ignored.

Examples

```
SHOW VERSION; -- will show KeySQL version
/*
 * Print the current user properties
 */
SHOW USER;
```

3.3   Elementary k-object data types

KeySQL currently supports four elementary k-object data types that are generally storage dependent. The elementary data types are described in the following table.

Data type	Description	Limitations
CHAR	String of variable length	Storage dependent
INTEGER	Signed 64-bit integer value	-9 223 372 036 854 775 808 .. 9 223 372 036 854 775 807

NUMBER	Value of any type of number	Approximately from 1E-307 to 1E+308
DATE	Date and time value	Storage dependent

The data types of the elementary k-objects in a catalog are specified in the respective CREATE KEYOBJECT statements.

3.4   Storage management statements

KeySQL database administrator can define a separate physical storage for any schema.

Default storage

Every KeySQL server instance has at least one storage named DEFAULT which is required to start the KeySQL server instance. The default storage has the following functions.

- It retains all metadata for a KeySQL instance.
- When the storage is not specified in a CREATE SCHEMA statement, the DEFAULT one is used.

> KeySQL server cannot create a schema without specifying an existing or default storage.

Connection strings

KeySQL uses the widespread connection string format common for ODBC, JDBC, .NET etc., as follows.

```
option_name_1 = option_value_1; option_name2 = option_value2; ...
```

where option_value_N is a scalar value, a string, or a composed value enclosed in braces {}.

Every option excluding Provider may be optional depending on the provider.

Option name	Description	Values
Provider	Type of underlying DBMS storage	Exemplary values: • Greenplum • MSSQL Server • Oracle • PostgreSQL • Teradata
Server	DBMS server name or address	valid network name or IP address
ServerPort	DBMS server port number	An integer number from 1 to 65536 Uses default port for selected DBMS when not specified
Database	Storage database name	
Schema	Storage default schema within the database	Used default schema when not specified. For example, public for PostgreSQL or dbo for MSSQL Server
UserName	DBMS connection login	
Password	DBMS connection password	
StorePassword	Specify whether the password is stored. Default is true. When true, the password is stored as encoded string	true or false
ConnectTimeout	Timeout in seconds to connect to DBMS. When not specified, the specific DBMS default value is used.	An integer number from 0 to 2 147 483 648 If 0, the connection will wait indefinitely.

The storage management statements are as follows.

CREATE STORAGE
ALTER STORAGE
DROP STORAGE
SHOW STORAGE
SHOW STORAGES
LIST STORAGES

CREATE CONNECTION_STRING
ALTER CONNECTION_STRING
DROP CONNECTION_STRING
SHOW CONNECTION_STRING
SHOW CONNECTION_STRINGS
LIST CONNECTION_STRING

CREATE STORAGE

The CREATE STORAGE statement creates a storage that will contain all underlying keyobject instances.

```
CREATE STORAGE <storage_name> [ WITH <comma_separated_option_list> ]

<option> ::=
    CONNECTION_STRING "="
        ( ' <connection_string> ' | <conn_string_name> ) [ NOCHECK ]
    | MAX_SIZE "=" ( <number> [ MB | GB ] | UNLIMITED )
    | MAX_INSTANCES "=" ( <int64_number> | UNLIMITED )
<storage_name> ::= <identifier>
<conn_string_name> ::= <identifier>
```

Options:

- `<storage_name>` must be unique within the KeySQL server instance.
- `CONNECTION_STRING` (mandatory) specifies a connection to the physical storage (data source), see Connection strings for more details.
    - `<conn_string_name>` is the name of existing connection string. See CREATE CONNECTION_STRING for more details.
    - NOCHECK (optional) directive indicates not to try to connect to the specified data source. This may be required when creating a storage on servers that are not running now. By default, the statement will try to connect to the specified data source and will return an error when connection fails.

- **MAX_SIZE** (optional) limits the storage size in the specified units: megabytes or gigabytes. The size is unlimited when not specified. Default units are the megabytes. Accepted values should be equal to or greater than 0.
- **MAX_INSTANCES** (optional) limits the total count of the storage instances. Unlimited when not specified.

> The zero (0) value for MAX_SIZE and MAX_INSTANCES options means that no user data may be inserted in the database. However, the user can still create schemas, catalogs, and keyobjects.

Permissions

The statement requires CREATE STORAGE permission.

Example

```
CREATE STORAGE staging
WITH CONNECTION_STRING = 'Provider=PostgreSQL; Server=localhost;
Database=staging; UserName=keysql; Password=gxT6jFh', MAX_SIZE =
UNLIMITED, MAX_INSTANCES = UNLIMITED
```

ALTER STORAGE

The ALTER STORAGE statement modifies the specified storage.

```
ALTER STORAGE <storage_name> WITH <comma_separated_option_list>

<option> ::=
    NAME "=" <new_storage_name>
    | CONNECTION_STRING "="
        ( ' <connection_string> ' | <conn_string_name> ) [ NOCHECK ]
    | MAX_SIZE "=" ( <number> [ MB | GB ] | UNLIMITED )
    | MAX_INSTANCES "=" ( <int64_number> | UNLIMITED )
```

In addition to the options of CREATE STORAGE mentioned above, the ALTER statement has following options:

- **NAME** will rename an existing storage to `<new_storage_name>`

> Note that changing a storage connection string starts the checking of the new physical storage. If the underlying structure does not match the storage metadata, an error is raised.

Permissions

The statement requires ALTER STORAGE permission.

Example

```
ALTER STORAGE staging WITH MAX_SIZE = 100 GB
```

DROP STORAGE

The DROP STORAGE statement deletes the specified storage.

```
DROP STORAGE <storage_name>
```

You cannot drop a storage assigned to some schema.

> ```
> The statement returns an error when the storage is in use by one
> or more KeySQL schemas.
> ```

Permissions

The statement requires DROP STORAGE permission.

Example

```
DROP STORAGE [ IF EXISTS ] staging
```

SHOW STORAGE

The SHOW STORAGE statement outputs the properties of the specified storage.

```
SHOW STORAGE <storage_name>
```

The statement also returns the inner keyobject used_by_schemas containing the list of names of schemas which use the specified storage.

In addition, the statement returns the inner keyobject storage_usage which contains the following information:

Property name	Caption	Values

used_size	The size in bytes of actually allocated storage space	Integer 0 .. 9223372036854775807
used_size_pretty	The size of allocated space in the human friendly format including size units starting from MB	nnnn.nn MB | GB | TB
max_size_pretty	Idem for the maximal storage size if specified	nnnn.nn MB | GB | TB or 'Unlimited'
used_size_percent	The percent of allocated space. Shows NULL when maximal size is not limited.	Decimal number with two digits to the right of the point 0.00 .. 100.00 or NULL

Permissions

The statement requires SHOW STORAGE permission.

Examples

```
SHOW STORAGE staging
```

Result:

```
{
    storage: {
        storage_id: 2,
        storage_name: 'STAGING',
        connection_string:
'Provider=PostgreSQL;Server=localhost;Database=staging;UserName=keysq
l;Password=***',
        max_size: 0,
        unlimited_size: 1,
        max_instances: 0,
        unlimited_instances: 1,
        owner_id: 3
        storage_usage: {
            used_size: 981467136,
```

```
            used_size_pretty: '0.91 GB',
            max_size_pretty: '1024.00 GB',
            used_size_percent: 0.09
        },
        used_by_schemas: {
            schema_name: 'test_acceptance',
            schema_name: 'test_coverage'
        }
    }
}
```

SHOW STORAGES

SHOW STORAGES statement returns the properties of all existing storages.

SHOW STORAGES

Permissions

The statement requires SHOW STORAGE permission.

Examples

SHOW STORAGES

Result:

```
{
    storage: {
        storage_id: 2,
        storage_name: 'STAGING',
        connection_string:
'Provider=PostgreSQL;Server=localhost;Database=staging;user=keysql;Pa
ssword=***',
        max_size: 0,
        unlimited_size: 1,
        max_instances: 0,
        unlimited_instances: 1,
        owner_id: 3
        storage_usage: {
            used_size: 981467136,
            used_size_pretty: '0.91 GB',
            max_size_pretty: '1024.00 GB',
            used_size_percent: 0.09
        },
        used_by_schemas: {
            schema_name: 'test_acceptance',
            schema_name: 'test_coverage'
        }
```

```
            }
    },
    {
        storage: {
            storage_id: 1,
            storage_name: 'DEFAULT',
            connection_string:
'Provider=PostgreSQL;Server=127.0.0.1;Database=keysql;Schema=keysql_d
ata;UserName=keysql;Password=***',
            max_size: 0,
            unlimited_size: 0,
            max_instances: 0,
            unlimited_instances: 0,
            owner_id: 1
            storage_usage: {
                used_size: 4189578,
                used_size_pretty: '4.00 MB',
                max_size_pretty: 'Unlimited',
                used_size_percent: NULL
            },
            used_by_schemas: {
                schema_name: 'PUBLIC',
                schema_name: 'SYSTEM',
                schema_name: 'KEYSQLUSER'
            }
        }
    },
    ...
```

LIST STORAGES

LIST STORAGES statement returns the lexicographically ordered list of all existing storages.

```
LIST STORAGES
```

Permissions

The statement requires LIST STORAGE permission.

Examples

```
LIST STORAGES
```

Result:

```
{storage_name:'DEFAULT'},
{storage_name:'STAGING'},
```

...

CREATE CONNECTION_STRING

The CREATE CONNECTION_STRING statement creates a named connection string (alias) which may be used by multiple storages.

The default connection string name is DEFAULT. You cannot create, alter, or drop the default connection string.

```
CREATE CONNECTION_STRING <connection_string_name> WITH
<comma_separated_option_list>

<option> ::=
    VALUE "=" ' <connection_string> ' [ NOCHECK ]
```

Options:

- `<connection_string_name>` must be unique within the KeySQL server instance.
- `VALUE` (mandatory) specifies a connection to the physical storage (data source), see Connection strings for more details.
- `NOCHECK` (optional) directive indicates not to try to connect to the specified data source.

Permissions

As connection strings are related to storages, the permissions for all operations with connection strings are the same.

The statement requires CREATE STORAGE permission.

Example

```
CREATE CONNECTION_STRING staging WITH VALUE = 'Provider=PostgreSQL;
Server=srv-test; Database=staging; UserName=keysql; Password=gxT6jFh'
```

ALTER CONNECTION_STRING

The ALTER CONNECTION_STRING statement modifies the specified connection string.

```
ALTER CONNECTION_STRING <connection_string_name>
    WITH <comma_separated_option_list>

<option> ::=
    NAME "=" <new_connection_string_name >
```

```
| VALUE "=" ' <connection_string> ' [ NOCHECK ]
```

In addition to the options of CREATE CONNECTION_STRING mentioned above, the ALTER statement has the following options:

- **NAME** will rename an existing connection string to `<new_storage_name>`

> Note that changing a connection string starts the checking of the new physical storage. If the underlying structure does not match the metadata of all related storages, an error is raised.

Permissions

The statement requires ALTER STORAGE permission.

Example

```
ALTER CONNECTION_STRING staging WITH NAME = staging_local
```

DROP CONNECTION_STRING

The DROP CONNECTION_STRING statement deletes the specified connection string.

```
DROP CONNECTION_STRING <connection_string_name>
```

You cannot drop a connection string in use by one or more storages.

> The statement returns an error when an existing storage uses the connection string.

Permissions

The statement requires DROP STORAGE permission.

Example

```
DROP CONNECTION_STRING staging_local
```

SHOW CONNECTION_STRING

The SHOW CONNECTION_STRING statement outputs the properties of the specified connection string.

```
SHOW CONNECTION_STRING <storage_name>
```

In addition, the statement returns the instance of keyobject used_by_storages containing the list of storages which use the specified connection string.

Permissions

The statement requires SHOW STORAGE permission.

Examples

```
SHOW CONNECTION_STRING staging_local
```

Result:

```
{
    conn_str: {
        conn_str_id: 2,
        conn_str_name: 'STAGING_LOCAL',
        conn_str_value: 'Provider=PostgreSQL;Server=srv-
test;ServerPort=5432;Database=keysql;UserName=keysql;Password=***;Con
nectTimeout=5',
        conn_str_nocheck: 0,
        owner_id: 3,
        used_by_storages: {
            storage_name:'STG_TEST_01',
            storage_name:'STG_TEST_02'
        }
    }
}
```

SHOW CONNECTION_STRINGS

SHOW CONNECTION_STRINGS statement returns the properties of all existing connection strings.

```
SHOW CONNECTION_STRINGS
```

Permissions

The statement requires SHOW STORAGE permission.

Examples

```
SHOW CONNECTION_STRINGS
```

Returns

```
{
    conn_str: {
        conn_str_id: 2,
        conn_str_name: 'STAGING_LOCAL',
        conn_str_value: 'Provider=PostgreSQL;Server=srv-
test;ServerPort=5432;Database=keysql;UserName=keysql;Password=***;Con
nectTimeout=5',
        conn_str_nocheck: 0,
        owner_id: 3,
        used_by_storages: {
            storage_name:'STG_TEST_01',
            storage_name:'STG_TEST_02'
        }
    }
},
{
    conn_str: {
        conn_str_id: 1,
        conn_str_name: 'DEFAULT',
        conn_str_value: 'Provider=PostgreSQL;Server=srv-
prod;ServerPort=5432;Database=keysql;Schema=keysql_data;UserName=keys
ql;Password=***;ConnectTimeout=5',
        conn_str_nocheck: 0,
        owner_id: 1,
        used_by_storages: {
            storage_name:'DEFAULT'
        }
    }
}
```

LIST CONNECTION_STRINGS

LIST CONNECTION_STRINGS statement returns the lexicographically ordered
list of all existing connection strings.

```
LIST CONNECTION_STRINGS
```

Permissions

The statement requires LIST STORAGE permission.

Examples

```
LIST CONNECTION_STRINGS
```

Result:

```
{conn_str_name: 'STAGING_LOCAL'},
```

```
{conn_str_name: 'DEFAULT'},
...
```

3.5   Schema management statements

The schema management statements are as follows.

CREATE SCHEMA
ALTER SCHEMA
DROP SCHEMA
SHOW SCHEMA
SHOW SCHEMAS
LIST SCHEMAS
DESCRIBE SCHEMA

CREATE SCHEMA

The CREATE SCHEMA statement creates a schema with the specified name that can contain catalogs and stores defined by the respective CREATE statements. All catalog and store names within a schema must be unique.

```
CREATE SCHEMA <schema_name> [ WITH <comma_separated_option_list> ]

<option> ::=
    STORAGE "=" <storage_name>
    | OWNER "=" <owner_name>
<owner_name> ::= <identifier>
```

Options:

- STORAGE (optional) specifies a storage name that will be used to store the data. If not specified, the default storage is used.
- OWNER (optional) specifies a username that will be assigned as the schema owner.

Permissions

The statement requires CREATE SCHEMA permission.

Example

```
CREATE SCHEMA MY_SCHEMA WITH STORAGE = staging
```

ALTER SCHEMA

The ALTER SCHEMA statement modifies the specified schema.

```
ALTER SCHEMA <schema_name> [ WITH <comma_separated_option_list> ]

<option> ::=
    NAME "=" <new_schema_name>
    | STORAGE "=" <storage_name>
    | OWNER "=" <owner_name>
```

Options:

- NAME renames the schema to <new_schema_name>
- STORAGE specifies a storage name that will be used to store the data. When the schema already contains one or more stores, you cannot change the storage.
- OWNER specifies a username that will become the schema owner.

Permissions

The statement requires ALTER SCHEMA permission.

Example

```
ALTER SCHEMA MY_SCHEMA WITH NAME = your_schema;
```

DROP SCHEMA

The DROP SCHEMA statement deletes the specified schema.

```
DROP SCHEMA [ IF EXISTS ] <schema_name>
```

You need to empty the schema to drop it by dropping all its catalogs.

> The statement returns an error when the schema contains catalogs.

Note that a catalog can be only dropped when there are no stores based on the catalog. So, the stores must be dropped first.

Permissions

The statement requires DROP SCHEMA permission.

Example

```
DROP SCHEMA MY_SCHEMA
```

SHOW SCHEMA

SHOW SCHEMA statement outputs the properties of the specified schema.

```
SHOW SCHEMA  [ <schema_name> ]
```

If the name is not specified, the statement outputs the current schema name.

The initial current schema is the default user schema. Use SET CURRENT SCHEMA statement to switch between schemas or display the name of the current one.

Permissions

The statement requires SHOW SCHEMA permission.

Examples

```
SHOW SCHEMA public
```

Result:

```
{
    SCHEMA_DESCRIPTION: {
        SCHEMA_NAME: 'PUBLIC',
        OWNER_NAME: 'keysqladmin',
        CATALOG_NAMES: {
            CATALOG_NAME: 'SK_CAT_BOM',
            CATALOG_NAME: 'TEMP_CAT',
            CATALOG_NAME: 'UNIVERSE'
        },
        STORE_NAMES: {
            STORE_NAME: 'BILLIARD_STORE'
        }
    }
}
```

SHOW SCHEMAS

SHOW SCHEMAS statement returns the properties of all existing schemas.

```
SHOW SCHEMAS
```

Permissions

The statement requires SHOW SCHEMA permission.

Examples

```
SHOW SCHEMAS
```

Result:

```
{
    SCHEMA_DESCRIPTION: {
        SCHEMA_NAME: 'PUBLIC',
        OWNER_NAME: 'keysqladmin',
        CATALOG_NAMES: {...},
        STORE_NAMES: {...}
    }
},
{
    SCHEMA_DESCRIPTION: {
        SCHEMA_NAME: 'SALES',
        OWNER_NAME: 'keysqladmin',
        CATALOG_NAMES: {...},
        STORE_NAMES: {...}
    }
},
...
```

LIST SCHEMAS

LIST SCHEMAS statement returns the lexicographically ordered list of all existing schemas.

```
LIST SCHEMAS
```

Permissions

The statement requires LIST SCHEMA permission.

Examples

```
LIST SCHEMAS
```

Result:

```
{SCHEMA_NAME: 'keysqluser'},
{SCHEMA_NAME: 'public'},
{SCHEMA_NAME: 'TEST'}
```

DESCRIBE SCHEMA

The DESCRIBE SCHEMA statement returns the KeySQL script to create the selected schema.

```
DESCRIBE SCHEMA <schema_name>
```

Additionally, the script may contain statements to create catalogs, keyobjects, and stores of the schema.

Permissions

The statement requires SHOW SCHEMA permission.

Examples

```
DESCRIBE SCHEMA public
```

Result:

```
{SCRIPT:'
--
-- schema PUBLIC
--
CREATE SCHEMA PUBLIC WITH STORAGE = DEFAULT, OWNER = j_smith;
--
-- catalog PUBLIC.UNIVERSE
--
CREATE CATALOG UNIVERSE;

-- catalog UNIVERSE, keyobjects:
CREATE KEYOBJECT YOB NUMBER IN CATALOG UNIVERSE;
CREATE KEYOBJECT RATING NUMBER IN CATALOG UNIVERSE;
...
CREATE KEYOBJECT LANDLORDS{LANDLORD,LANDLORD_P} IN CATALOG UNIVERSE;

-- store PUBLIC.OFFICE_RENTALS
CREATE STORE OFFICE_RENTALS FOR CATALOG UNIVERSE;'
}
```

3.6   Catalog management statements

A catalog is a set of k-object definitions. The catalog management statements are as follows.

CREATE CATALOG

ALTER CATALOG
DROP CATALOG
DROP CATALOGS
SHOW CATALOG
LIST CATALOGS
DESCRIBE CATALOG

CREATE KEYOBJECT
ALTER KEYOBJECT
DROP KEYOBJECT
SHOW KEYOBJECT
SHOW KEYOBJECTS
LIST KEYOBJECTS
DESCRIBE KEYOBJECT

CREATE SEQUENCE
ALTER SEQUENCE
DROP SEQUENCE
SHOW SEQUENCE
SHOW SEQUENCES
LIST SEQUENCES
DESCRIBE SEQUENCE

CREATE CATALOG

The CREATE CATALOG statement creates a catalog with the specified name in the specified or current schema.

```
CREATE CATALOG <qualified_catalog_name> [ WITH
<comma_separated_option_list> ]

<qualified_catalog_name> ::= [ <schema_name> "." ] <catalog_name>
<option> ::=
    OWNER "=" <owner_name>
```

When <schema_name> is not specified, the current schema is used.

Permission

The statement requires CREATE CATALOG permission.

Example

```
CREATE CATALOG MY_SCHEMA.MY_CATALOG
```

ALTER CATALOG

The ALTER CATALOG statement modifies the properties of an existing catalog.

```
ALTER CATALOG <qualified_catalog_name> [ WITH
<comma_separated_option_list> ]

<option> ::=
    NAME "=" <new_catalog_name>
    | OWNER "=" <owner_name>
```

Options

- NAME renames the catalog within the schema to <new_catalog_name>
- OWNER specifies a username that will become the owner of specified catalog.

Permissions

The statement requires ALTER CATALOG permission.

Example

```
ALTER CATALOG MY_SCHEMA.MY_CATALOG WITH NAME = sales_2020, OWNER =
jsmith
```

DROP CATALOG

The DROP CATALOG statement deletes the specified catalog.

```
DROP CATALOG [ IF EXISTS ] <qualified_catalog_name>
```

The specified catalog should not be related to the existing stores.

> The statement returns an error when a related store exists.

Permissions

The statement requires DROP CATALOG permission.

Example

```
DROP CATALOG MY_SCHEMA.MY_CATALOG
```

DROP CATALOGS

The DROP CATALOGS statements delete all catalogs in the specified schema if all related stores had been already dropped.

```
DROP CATALOGS FROM SCHEMA <schema_name>
```

Permissions

The statement requires DROP CATALOG permission.

Example

```
DROP CATALOGS FROM my_schema
```

SHOW CATALOG

The statement returns the header identifying the catalog followed by the comma separated and lexicographically ordered list of definitions for all its k-objects enclosed in braces.

```
SHOW CATALOG <qualified_catalog_name>
```

Permissions

The statement requires SHOW CATALOG permission.

Example

```
SHOW CATALOG public.universe
```

The output:

```
{
    CATALOG_DESCRIPTION : {
        SCHEMA_NAME : 'PUBLIC',
        CATALOG_NAME : 'UNIVERSE',
        OWNER_NAME : 'keysqladmin',
        STORE_NAMES : {
            STORE_NAME : 'BILLIARD_STORE',
        },
        KEYOBJECT_DEFINITIONS : {
            KEYOBJECT_DEFINITION : {
                KEYOBJECT_NAME : 'UPC',
                KEYOBJECT_TYPE : 'NUMBER',
                COMPONENT_KEYOBJECT_NAMES : NULL,
            },
```

```
    KEYOBJECT_DEFINITION : {
        KEYOBJECT_NAME : 'BRAND',
        KEYOBJECT_TYPE : 'CHAR',
        COMPONENT_KEYOBJECT_NAMES : NULL,
    }
...
}
```

SHOW CATALOGS

The SHOW CATALOGS statement returns the properties of all existing catalogs within the specified schema.

```
SHOW CATALOGS [ [ FROM [ SCHEMA ] <schema_name> ] | ALL ]
```

Options:

- FROM [SCHEMA] <schema_name> - show catalogs from specified schema only. The SCHEMA keyword is optional.
- ALL – show all catalogs from all schemas.

When the schema name or the ALL keyword are not specified, the current schema is used.

Permissions

The statement requires SHOW CATALOG permission.

Examples

```
SHOW CATALOGS FROM sales_2020
```

Result

```
{
    CATALOG_DESCRIPTION : {
        SCHEMA_NAME : 'PUBLIC',
        CATALOG_NAME : 'orders',
        OWNER_NAME : 'keysqladmin',
        STORE_NAMES : {...},
        KEYOBJECT_DEFINITIONS : {...}
    ...
},
{
    CATALOG_DESCRIPTION : {
        SCHEMA_NAME : 'PUBLIC',
        CATALOG_NAME : 'sales',
        OWNER_NAME : 'keysqladmin',
        STORE_NAMES : {...},
```

```
        KEYOBJECT_DEFINITIONS : {...}
    ...
}
```

Other examples

```
SHOW CATALOGS FROM SCHEMA sales_2020;
SHOW CATALOGS;
SHOW CATALOGS ALL;
```

LIST CATALOGS

The LIST CATALOGS statement returns the lexicographically ordered list of all existing catalogs within the specified schema.

```
LIST CATALOGS
    [ FROM [ SCHEMA ] <schema_name> | ALL ]
```

Options:

- FROM [SCHEMA] <schema_name> - list catalogs from specified schema only. The SCHEMA keyword is optional.
- ALL – list all catalogs from all schemas.

When the schema name or the ALL keyword are not specified, the current schema is used.

Permissions

The statement requires LIST CATALOG permission.

Examples

```
LIST CATALOGS sales_2020
```

Result:

```
{CATALOG_NAME:'orders'},
{CATALOG_NAME:'sales'}
```

DESCRIBE CATALOG

The DESCRIBE CATALOG statement returns the KeySQL script to create the selected catalog.

```
DESCRIBE CATALOG <qualified_catalog_name>
```

When schema name is not specified, the current schema is used.

Additionally, the script may contain statements to create keyobjects of the catalog.

Permissions

The statement requires SHOW CATALOG permission.

Examples

```
DESCRIBE CATALOG public.universe
```

Result:

```
{SCRIPT:'
--
-- catalog PUBLIC.UNIVERSE
--
CREATE CATALOG UNIVERSE;

-- catalog UNIVERSE, keyobjects:
CREATE KEYOBJECT YOB NUMBER IN CATALOG UNIVERSE;
CREATE KEYOBJECT RATING NUMBER IN CATALOG UNIVERSE;
...
CREATE KEYOBJECT LANDLORDS{LANDLORD,LANDLORD_P} IN CATALOG UNIVERSE;'
}
```

CREATE KEYOBJECT

The CREATE KEYOBJECT statement creates a new k-object with the specified name and type. The type can be elementary, a composition, or a multi-composition.

```
CREATE KEYOBJECT <keyobject_name> <keyobject_type>
    IN [ CATALOG ] <qualified_catalog_name>

<keyobject_type> ::= <elementary_type> | <composition> | <multi-
composition>
<elementary_type> ::= CHAR | INTEGER | NUMBER | DATE
<composition> ::= "{" <list_of_keyobject_names> "}"
<multi-composition> ::= "{" <keyobject name> MULTIPLE "}"
<list_of_keyobject_names> ::= <keyobject name> [ ","
<list_of_keyobject_names> ]
```

The elementary types are described above. See "Elementary data types" for details.

K-objects referenced in the composition or multi-composition definitions must already exist in the same catalog.

Permissions

The statement requires CREATE KEYOBJECT permission (catalog scope).

Examples

```
CREATE KEYOBJECT AGE NUMBER IN CATALOG MY_SCHEMA.MY_CATALOG;
CREATE KEYOBJECT FIRST_NAME CHAR IN CATALOG MY_SCHEMA.MY_CATALOG;
CREATE KEYOBJECT LAST_NAME CHAR IN CATALOG MY_SCHEMA.MY_CATALOG;
CREATE KEYOBJECT PERSON {LAST_NAME, FIRST_NAME} IN
MY_SCHEMA.MY_CATALOG;
CREATE KEYOBJECT PEOPLE {PERSON MULTIPLE} IN CATALOG
MY_SCHEMA.MY_CATALOG;
```

ALTER KEYOBJECT

The ALTER KEYOBJECT statement alters existing k-object to a new name and/or type within the same catalog.

```
ALTER KEYOBJECT <keyobject_new_name> "=" <keyobject_name>
<keyobject_type>
  IN [ CATALOG ] <qualified_catalog_name>
```

The syntax allows to change both the name and the type of specified keyobject.

> The following syntax will be available in the future versions of KeySQL.

```
ALTER KEYOBJECT <keyobject_name>
  IN CATALOG <qualified_catalog_name>
  WITH <comma_separated_option_list>

<option> ::=
    NAME "=" <new_keyobject_name>
  | TYPE "=" <new_keyobject_type>
```

When altering just the name of a k-object (renaming a k-object), the type must be omitted.

Currently, the k-object type cannot be altered if the k-object has instances inserted in at least one store; a k-object can only be renamed. However, when the statement changes just the order of elements in a composite k-object, it does not require checking if the instances are already inserted.

Options:

- `NAME` renames the k-object to `<new_keyobject_name>`
- `TYPE` specifies a new data type.

Permissions

The statement requires ALTER KEYOBJECT permission (catalog scope).

Examples

```
-- rename only
ALTER KEYOBJECT last_name = l_name CHAR IN CATALOG public.contacts;

-- change type
ALTER KEYOBJECT person = person { last_name, first_name,
date_of_birth }
  IN CATALOG public.contacts;
ALTER KEYOBJECT person_id = person_id CHAR IN CATALOG
public.contacts;
```

> The following syntax will be available in the future versions of KeySQL.

```
-- rename only
ALTER KEYOBJECT l_name
  IN CATALOG public.contacts
  WITH NAME = last_name;

-- change type
ALTER KEYOBJECT person
  IN CATALOG public.contacts
  WITH TYPE = { last_name, first_name, date_of_birth };
ALTER KEYOBJECT person_id
  IN CATALOG public.contacts
  WITH TYPE = CHAR;
```

DROP KEYOBJECT

The DROP KEYOBJECT statement removes the k-object from the catalog.

```
DROP KEYOBJECT [ IF EXISTS ] <keyobject name>
FROM [ CATALOG ] <qualified_catalog_name>
```

A k-object can only be dropped if it is not referenced by other k-objects and there are no instances of the k-object in any store based on the catalog.

Permissions

The statement requires DROP KEYOBJECT permission (catalog scope).

Examples

```
DROP KEYOBJECT PERSON FROM CATALOG MY_SCHEMA.MY_CATALOG;
DROP KEYOBJECT LAST_NAME FROM CATALOG MY_SCHEMA.MY_CATALOG;
/* Note the composite keyobject PERSON being dropped first */
```

SHOW KEYOBJECT

The SHOW KEYOBJECT statement returns the definition of the k-object in the catalog.

```
SHOW KEYOBJECT <keyobject name>
FROM [ CATALOG ] <qualified_catalog_name>
```

Permissions

The statement requires SHOW KEYOBJECT permission (catalog scope).

Examples

```
SHOW KEYOBJECT brand FROM public.universe
```

Result:

```
{
  KEYOBJECT_DESCRIPTION : {
    CATALOG_NAME : 'UNIVERSE',
    KEYOBJECT_DEFINITION : {
      KEYOBJECT_NAME : 'BRAND',
      KEYOBJECT_TYPE : 'CHAR',
      COMPONENT_KEYOBJECT_NAMES : NULL
    },
    COMPONENT_KEYOBJECTS : NULL
  }
}
```

SHOW KEYOBJECTS

The SHOW KEYOBJECTS statement returns the definition of the k-objects in the catalog.

```
SHOW KEYOBJECTS FROM [ CATALOG ] <qualified_catalog_name>
```

Permissions

The statement requires SHOW KEYOBJECT permission (catalog scope).

Examples

```
SHOW KEYOBJECTS FROM public.universe
```

Result:

```
{
  KEYOBJECT_DESCRIPTION : {
    CATALOG_NAME : 'UNIVERSE',
    KEYOBJECT_DEFINITION : {
      KEYOBJECT_NAME : 'BRAND',
      KEYOBJECT_TYPE : 'CHAR',
      COMPONENT_KEYOBJECT_NAMES : NULL
    },
    COMPONENT_KEYOBJECTS : NULL
  },
  ...
  KEYOBJECT_DESCRIPTION: {
      CATALOG_NAME: 'UNIVERSE',
      KEYOBJECT_DEFINITION: {
          KEYOBJECT_NAME: 'BALL',
          KEYOBJECT_TYPE: 'COMPOSITION',
          COMPONENT_KEYOBJECT_NAMES: {
              KEYOBJECT_NAME: 'NUM',
              KEYOBJECT_NAME: 'COLOR'
          }
      },
      COMPONENT_KEYOBJECTS: {
          KEYOBJECT_DEFINITION: {
              KEYOBJECT_NAME: 'NUM',
              KEYOBJECT_TYPE: 'NUMBER',
              COMPONENT_KEYOBJECT_NAMES: NULL
          },
          KEYOBJECT_DEFINITION: {
              KEYOBJECT_NAME: 'COLOR',
              KEYOBJECT_TYPE: 'CHAR',
              COMPONENT_KEYOBJECT_NAMES: NULL
          }
      }
  },
  ...
}
```

LIST KEYOBJECTS

The LIST KEYOBJECTS statement returns the lexicographically ordered list of all k-objects of the specified catalog.

```
LIST KEYOBJECTS
   FROM [ CATALOG ] <qualified_catalog_name>
```

Permissions

The statement requires LIST KEYOBJECTS permission (catalog scope).

Examples

```
LIST KEYOBJECTS FROM public.universe
```

Result:

```
{ KEYOBJECT_NAME : 'AGGR_RESULT' },
{ KEYOBJECT_NAME : 'AVG_LENGTH' },
{ KEYOBJECT_NAME : 'BALL' },
...
```

DESCRIBE KEYOBJECT

The DESCRIBE KEYOBJECT statement returns the KeySQL script to create the specified keyobject.

```
DESCRIBE KEYOBJECT <keyobject_name> FROM [ CATALOG ]
<qualified_catalog_name>
```

When schema name is not specified, the current schema is used.

Permissions

The statement requires SHOW KEYOBJECT permission.

Examples

```
DESCRIBE KEYOBJECT person FROM universe
```

Result:

```
{
  SCRIPT : '
CREATE KEYOBJECT PERSON{FIRST_NAME,LAST_NAME,YOB} IN CATALOG
UNIVERSE;'
}
```

CREATE SEQUENCE

A sequence is a user-defined, catalog-bound, integer-type k-object that generates a sequence of values according to the specification.

```
CREATE SEQUENCE <sequence_name>
    IN [ CATALOG ] <qualified_catalog_name>
    [ WITH <comma_separated_option_list> ]

<sequence_name> ::= <identifier>
<option> ::=
    MIN "=" <value_int64>
    | MAX "=" <value_int64>
    | INITIAL "=" <value_int64>
    | INCREMENT "=" <value_int32>
```

Options:

- MIN specifies the lower bound for the sequence k-object. The parameter is optional without the default value.
- MAX specifies the upper bound for the sequence k-object. The parameter is optional without the default value.
- INITIAL specifies the first value returned by the sequence. The parameter is optional, the default value is 1. The value must be less than or equal to the specified maximum value and greater than or equal to the specified minimum value, otherwise KeySQL returns an error.
- INCREMENT value used to increment (or decrement if negative) the current value of the sequence for each usage. The parameter is optional, the default value is 1.

Permissions

The statement requires CREATE KEYOBJECT permission (catalog scope).

Examples

```
CREATE CATALOG public.employees;
CREATE SEQUENCE seq_emp IN CATALOG public.employees WITH MIN = 1,
INCREMENT = 1;
CREATE SEQUENCE seq_jobs IN CATALOG public.employees WITH MIN = 1,
INCREMENT = 1;
```

ALTER SEQUENCE

The ALTER SEQUENCE statement modifies properties of an existing sequence k-object.

```
ALTER SEQUENCE <sequence_name>
    IN [ CATALOG ] <qualified_catalog_name>
    [ WITH <comma_separated_option_list> ]

<option> ::=
    MIN "=" <value_int64>
    | MAX "=" <value_int64>
    | INITIAL "=" <value_int64>
    | INCREMENT "=" <value_int32>
```

Every modification of the initial value will reset the sequence.

DESCRIBE SEQUENCE

The DESCRIBE SEQUENCE statement returns the KeySQL script to create the specified sequence.

```
DESCRIBE SEQUENCE <sequence_name> FROM <qualified_catalog_name>
```

When schema name is not specified, the current schema is used.

Permissions

The statement requires SHOW CATALOG permission.

Examples

```
-- CREATE SEQUENCE id_seq IN CATALOG universe
DESCRIBE SEQUENCE id_seq FROM CATALOG public.universe
```

Result:

```
{
  SCRIPT : '
CREATE SEQUENCE ID_SEQ IN CATALOG UNIVERSE WITH INITIAL = 1,
INCREMENT = 1, MIN = 1, MAX = 9223372036854775807;'
}
```

Permissions

The statement requires ALTER KEYOBJECT permission for the catalog.

Example

```
ALTER SEQUENCE seq_emp IN public.employees WITH INCREMENT = 10
```

DROP SEQUENCE

DROP SEQUENCE statement deletes the specified sequence k-object. If the sequence k-object is used by one or more constraints, the statement returns an error.

```
DROP SEQUENCE [ IF EXISTS ] <sequence_name>
    FROM [ CATALOG ] <qualified_catalog_name>
```

Permissions

The statement requires DROP KEYOBJECT permission (catalog scope).

Example

```
DROP SEQUENCE seq_emp FROM public.employees
```

SHOW SEQUENCE

SHOW SEQUENCE statement returns all properties of an existing sequence k-object.

```
SHOW SEQUENCE <sequence_name>
    FROM [ CATALOG ] <qualified_catalog_name>
```

Permissions

The statement requires SHOW KEYOBJECT permission (catalog scope).

Example

```
SHOW SEQUENCE seq_employee FROM hr
```

The output

```
{
  SEQUENCE_INFO : {
    CATALOG_NAME : 'HR',
    SEQUENCE_DEFINITION : {
      SEQUENCE_NAME : 'SEQ_EMPLOYEE',
      SEQUENCE_OPTIONS : {
        INITIAL_VALUE : 1,
        MIN_VALUE : 1,
        MAX_VALUE : 9223372036854775807,
        INCREMENT : 1
      }
    },
```

```
      SEQUENCE_REFS : {
         SEQUENCE_REF : {
            STORE_NAME : 'EMPLOYEES',
            CONSTRAINT : {
               CONSTRAINT_NAME : 'SEQUENCE_EMP_NO',
               CONSTRAINT_DEFINITION :
'DEFAULT(EMPLOYEE.EMP_NO,NEXTVAL(SEQ_EMPLOYEE))'
            }
         }
      }
   }
}
```

SHOW SEQUENCES

SHOW SEQUENCES statement returns the properties of all sequence k-objects in the specified catalog.

```
SHOW SEQUENCES
   FROM [ CATALOG ] <qualified_catalog_name>
```

Permissions

The statement requires SHOW KEYOBJECT permission (catalog scope).

Example

```
SHOW SEQUENCES FROM employees
```

LIST SEQUENCES

The LIST SEQUENCES statement returns the lexicographically ordered list of sequence names for the specified catalog.

```
LIST SEQUENCES
   FROM [ CATALOG ] <qualified_catalog_name>
```

Permissions

The statement requires LIST KEYOBJECT permission (catalog scope)

Example

```
LIST SEQUENCES FROM hr
```

The output:

```
{
  SEQUENCE_NAME : 'SEQ_EMPLOYEE'
}
```

NEXTVAL function

The NEXTVAL function returns the `INITIAL` value on the first use of the sequence. Subsequently, it increments the current value with the `INCREMENT` amount, and returns the result. If the incremented `INITIAL` value is not between the `MIN` and the `MAX`, the function returns an error.

```
NEXTVAL "(" <sequence_name> "," [ <qualified_catalog_name> ] ")"
```

The <qualified_catalog_name> may be optional when NEXTVAL function is used in the context of a store constraint, i.e., in DEFAULT expression.

Permissions

None.

When used in the context of SELECT, INSERT or UPDATE statements, the function usage requires corresponding permission for the store where the sequence is used.

Examples

Consider the catalog definition from the Store constraints.

Example

```
SELECT NEXTVAL(seq_employee, hr);
```

Result:

```
{ NEXTVAL: 1 }
```

Using the sequence in DEFAULT constraint within INSERT statement.

```
INSERT INTO employees INSTANCES
-- use DEFAULT constraint for emp_no
{
    employee: {first_name: 'John', last_name: 'Smith', age: 35,
      emails: {email: 'js@net.co'},
      children: {child: {first_name: 'Nancy', last_name: 'Smith',
age: 12}},
      trainings: {training: {succeeded: 'Y'}}
  }
```

```
},
-- use DEFAULT keyword as value for emp_no
{
    employee: {emp_no: DEFAULT, first_name: 'Jack', last_name:
'Sparrow', age: 27,
        emails: {email: 'jsp@net.co'},
        children: {child: {first_name: 'Kate', last_name: 'Sparrow',
age: 3}},
        trainings: {training: {succeeded: 'Y'}}
    }
};
SELECT * FROM employees;
```

This statement returns:

```
{
  EMPLOYEE : {
    EMP_NO : 1,
    AGE : 35,
    HIRED_AT : '2021-06-30 14:48:01.090422',
    FIRST_NAME : 'John',
    LAST_NAME : 'Smith',
    EMAILS : {
      EMAIL : 'js@net.co'
    },
    CHILDREN : {
      CHILD : {
        AGE : 12,
        FIRST_NAME : 'Nancy',
        LAST_NAME : 'Smith'
      }
    },
    TRAININGS : {
      TRAINING : {
        NAME : NULL,
        SUCCEEDED : 'Y'
      }
    }
  }
},
{
  EMPLOYEE : {
    EMP_NO : 2,
    AGE : 27,
    HIRED_AT : '2021-06-30 14:48:01.090422',
    FIRST_NAME : 'Jack',
    LAST_NAME : 'Sparrow',
    EMAILS : {
      EMAIL : 'jsp@net.co'
```

```
    },
    CHILDREN : {
      CHILD : {
        AGE : 3,
        FIRST_NAME : 'Kate',
        LAST_NAME : 'Sparrow'
      }
    },
    TRAININGS : {
      TRAINING : {
        NAME : NULL,
        SUCCEEDED : 'Y'
      }
    }
  }
}
```

You may also explicitly specify NEXTVAL in the query without default constraints:

```
CREATE STORE emp_no_test FOR CATALOG hr
;
INSERT INTO emp_no_test
SELECT { NEXTVAL(seq_employee, hr) } AS emp_no;
```

Then check the store:

```
SELECT * FROM emp_no_test
```

The statement returns:

```
{
  EMP_NO: 3
}
```

3.7   Store management statements

The store management statements are as follows.

CREATE STORE
CREATE STORE AS
ALTER STORE
DROP STORE
TRUNCATE STORE

DROP STORES
SHOW STORE
LIST STORES
DESCRIBE STORE

Store hosts

A store host is a k-object the instances of which are inserted into the store. Server checks the hosts in the following cases:

- Inserting instances into the store using INSERT or INSERT SELECT statements. KeySQL returns an error when k-object is not allowed for the store.
- Modifying the store hosts with ALTER STORE statement.
- Selecting instances from the store using the SELECT statement. When the store contains a selected k-object, KeySQL server returns a non-empty or empty dataset, otherwise it returns an error.
- When a store is defined with no constraints on the hosts, the instances of any keyobject from the respective catalog can be inserted into the store.

Store constraints

Store constraints are used to specify rules for data in a store. Constraints can be used to restrict values of instances of certain k-objects within the hosts. This ensures the accuracy and reliability of the data. When a data modification action violates a constraint, the action is aborted.

Store constraints may be specified only for stores which have one or more explicitly declared hosts. The hosts can be declared as a part of the CREATE STORE or ALTER STORE statement. A store without the declared hosts is called *unconstrained.*

When specified, a constraint name must be unique within the store scope. The constraint name is optional, KeySQL server will assign an internal unique name if it is not explicitly specified.

The following types of constraints are supported.

- The UNIQUE constraint defines the set of elementary k-objects which values should be unique within the store host instances.
- The CHECK constraint defines the expression that will be verified every time an elementary k-object instance is inserted or changed.

- The multi-composition constraint specifies whether the instances comprise the set or the bag (multiset):
  - IS_SET: multi-composition k-object instance cannot contain duplicate k-object instances as elements.
  - IS_BAG (value by default): multi-composition k-object instance may contain duplicate k-object instances as elements.
- The nullability constraint NULL / NOT NULL allows or forbids the NULL values for elementary k-object instances. By default, NULL values are allowed.
- The DEFAULT constraint specifies a default value that will be assigned to the elementary k-object instance. The constraint can reference an existing sequence k-object.

CREATE STORE

The CREATE STORE statement creates a new store that can contain instances of any k-object from the catalog. Currently, store names are unique across all catalogs within a schema.

```
CREATE STORE <store_name>
    FOR [CATALOG] <qualified_catalog_name>
    [ WITH <comma_separated_option_list> ]

<option> ::=
      OWNER = <owner_name>
    | HOSTS "=" "{" <host_list> "}"
    | CONSTRAINTS "=" "{" <store_constraint_list> "}"

<host_list> ::= <host_name_list>
<host_name_list> ::= <host_name> [ "," <host_name_list> ]

<store_constraint_list> ::= <store_constraint> [
<store_constraint_list> ]
<store_constraint> ::=
  [ <constraint_name> ] <constraint_definition>

<constraint_name> ::= <identifier>
<constraint_definition> ::=
    <unique_constraint>
    | <multi_composition_constraint>
    | <nullability_constraint>
    | <default_constraint>
    | <check_constraint>

<unique_constraint> ::=
  UNIQUE "(" <elementary_keyobject_list> ")"
<elementary_keyobject_list> ::=
```

```
    <qualified_elementary_keyobject_name> [ ","
<elementary_keyobject_list> ]
<qualified_elementary_keyobject_name> ::=
    [ <qualified_catalog_name> "." ] <elementary_keyobject_name>

<multi_composition_constraint> ::=
    IS_BAG | IS_SET "(" <qualified_multi_keyobject_name> ")"

<nullability_constraint> ::=
    NULL | NOT_NULL "(" <qualified_elementary_keyobject_name> ")"

<default_constraint> ::=
    DEFAULT "(" <qualified_elementary_keyobject_name> ","
        <constant_expression> | NEXTVAL "(" <sequence_name> ")" ")"

<check_constraint> ::=
    CHECK "(" <logical_expresson> ")"
```

Where <logical_expresson> is any logical expression using elementary k-objects.

Options:

- OWNER specifies a username that will become the store owner.
- HOSTS lists the allowed host k-objects from the catalog.
- CONSTRAINTS lists the store constraints.

Permissions

The statement requires CREATE STORE permission.

Example

Create a store without constraints:

```
CREATE STORE blrd_store FOR CATALOG public.universe
```

Create a store with owner, hosts, and constraints:

```
CREATE CATALOG hr;
CREATE KEYOBJECT emp_no INTEGER IN CATALOG hr;
CREATE KEYOBJECT age INTEGER IN CATALOG hr;
CREATE KEYOBJECT hired_at DATE IN CATALOG hr;
CREATE KEYOBJECT first_name CHAR IN CATALOG hr;
CREATE KEYOBJECT last_name CHAR IN CATALOG hr;
CREATE KEYOBJECT name CHAR IN CATALOG hr;
CREATE KEYOBJECT succeeded CHAR IN CATALOG hr;
CREATE KEYOBJECT email CHAR IN CATALOG hr;
CREATE KEYOBJECT emails {email MULTIPLE} IN CATALOG hr;
```

```
CREATE KEYOBJECT child {first_name, last_name, age} IN CATALOG hr;
CREATE KEYOBJECT children {child MULTIPLE} IN CATALOG hr;
CREATE KEYOBJECT training {name, succeeded} IN CATALOG hr;
CREATE KEYOBJECT trainings {training MULTIPLE} IN CATALOG hr;
CREATE KEYOBJECT employee {emp_no, first_name, last_name, age,
hired_at,emails, children, trainings} IN CATALOG hr;
CREATE SEQUENCE seq_employee IN CATALOG hr;

CREATE STORE employees FOR CATALOG hr
WITH
   HOSTS = {employee},
   CONSTRAINTS = {
      CHECK(employee.age >= 18),
      CHECK(employee.children.age <= 18),
      chk_emp_training CHECK(employee.trainings.succeeded = 'Y' OR
                             employee.trainings.succeeded = 'N'),
      NOT_NULL (employee.children.first_name),
      DEFAULT (employee.hired_at, CURRENT_DATE()),
      uc_emp_email UNIQUE (employee.emails.email),
      sequence_emp_no DEFAULT (employee.emp_no,
NEXTVAL(seq_employee))
   }
;
```

CREATE STORE AS

The CREATE STORE AS statement creates a new store for the specified catalog that will contain all k-object instances returned by the respective SELECT statement. All stores and k-objects referenced in the SELECT statement must belong to the same catalog.

Currently, store names are unique across all catalogs within a schema.

```
CREATE STORE <store_name>
   FOR [CATALOG] <qualified_catalog_name>
   AS <select statement>
```

Permissions

The statement requires CREATE STORE permission.

Examples

```
CREATE STORE temp FOR CATALOG public.my_catalog AS
SELECT *
FROM my_store;
```

```
CREATE STORE some_numbers FOR blrd_catalog AS
SELECT DISTINCT X.NUM AS NUM
FROM BLRD_STORE X JOIN BLRD_STORE Y ON X.GAME = Y.GAME
WHERE Y.COLOR = 'red'
```

ALTER STORE

The ALTER STORE statement allows renaming a store and modifying its properties and constraints.

```
ALTER STORE <qualified_store_name> [ WITH
<comma_separated_option_list> ]

<qualified_store_name> ::= [ <schema_name> "." ] <store_name>
<option> ::=
    NAME "=" <new_store_name>
    | OWNER "=" <owner_name>
    | HOSTS "=" "{" <host_list> "}"
    | CONSTRAINTS "=" "{" <store_constraint_list> "}"
```

Options

- NAME renames the store within the same schema to <new_store_name>.
- OWNER specifies a username that will become the store owner.
- HOSTS lists the allowed host k-objects from the catalog.
- CONSTRAINTS lists the store constraints.

All options except NAME are the same as in the CREATE STORE statement.

The additional ALTER STORE syntax allows to add or drop the hosts and constraints from the specified store as follows.

```
ALTER STORE <qualified_store_name> ( ADD | DROP )
    HOSTS "{" <host_list> "}"
    | CONSTRAINTS "{" <store_constraint_list> "}"
```

During the execution of the ALTER STORE statements KeySQL server performs the run-time checks as follows:

- If the current hosts contradict the newly declared hosts, an error is raised, and the action is aborted.

- If the current data in the store contradicts a newly specified constraint, an error is raised, and the action is aborted.

Permissions

The statement requires ALTER STORE permission.

Examples

Rename the store and change its owner:

```
ALTER STORE hr.employees WITH NAME = your_name, OWNER = jsmith
```

Modify hosts:

```
ALTER STORE hr.employees DROP HOSTS {salaries};
ALTER STORE hr.employees ADD HOSTS {employee, workplaces};
```

Change the whole host list at once: drop all old values then add new ones:

```
ALTER STORE hr.employees WITH HOSTS = {employee, workplaces};
```

Add constraints:

```
ALTER STORE hr.employees ADD CONSTRAINTS {
   CHECK(employee.age >= 14 AND employee.age <= 99),
   CHECK(employee.children.age > 0 AND employee.children.age <= 18),
   chk_emp_training CHECK(employee.trainings.succeeded = 'Y' OR
                          employee.trainings.succeeded = 'N')
};

ALTER STORE hr.employees ADD CONSTRAINTS {
  NOT_NULL (employee.children.first_name),
  dft_emp_hired DEFAULT (employee.hired_at, CURRENT_DATE() )
};
```

Drop existing constraints:

```
ALTER STORE hr.employees DROP CONSTRAINTS {dft_emp_hired,
chk_emp_training};
```

Change the whole constraint list at once:

```
ALTER STORE hr.employees WITH CONSTRAINTS = {
   -- E-mail k-object uniqueness inside the collection only
   IS_SET(employee.emails),
   -- E-mail value (char) uniqueness across the whole store
   uc_emp_email UNIQUE (employee.emails.email)
};
```

Use a sequence:

```
ALTER STORE hr.employees ADD CONSTRAINT {
  dft_seq_emp DEFAULT (employee.emp_no, NEXTVAL(seq_employee))
};
```

DROP STORE

The DROP STORE statement drops the store with all k-object instances it contains.

```
DROP STORE [ IF EXISTS ] <qualified_store_name>
```

Permissions

The statement requires DROP STORE permission.

Example

```
DROP STORE public.blrd_store
```

TRUNCATE STORE

The statement deletes all instances of k-objects in the specified store. The DELETE statement without WHERE does the same but TRUNCATE is generally faster.

```
TRUNCATE [ STORE ] <qualified_store_name>
```

Permissions

The statement requires DELETE STORE permission.

Example

```
TRUNCATE public.billiard_store
```

DROP STORES

The DROP STORES statement drops all stores from the specified schema, or all stores related to the specified catalog.

```
DROP STORES FROM SCHEMA <schema_name>
DROP STORES FROM CATALOG <qualified_catalog_name>
```

Permissions

The statement requires DROP STORE permission.

Examples

```
DROP STORES FROM SCHEMA public;
DROP STORES FROM CATALOG public.universe;
```

SHOW STORE

The statement returns the header identifying the store and the catalog it is defined for, followed by the comma separated and lexicographically ordered list of definitions of all host k-objects of the store enclosed in braces.

```
SHOW STORE <qualified_store_name>
```

Permissions

The statement requires SHOW STORE permission.

Example

```
SHOW STORE public.billiard_store
```

The output:

```
{
  STORE_INFO : {
    STORE_NAME : 'BILLIARD_STORE',
    SCHEMA_NAME : 'PUBLIC',
    OWNER_NAME : 'keysqladmin',
    IS_EXTERNAL : 0,
    HOSTS : {
      KEYOBJECT_DEFINITION : {
        COMPONENT_KEYOBJECT_NAMES : {
          KEYOBJECT_NAME : 'BALLSET',
          KEYOBJECT_NAME : 'BRAND',
          KEYOBJECT_NAME : 'NAME',
          KEYOBJECT_NAME : 'UPC'
        },
        KEYOBJECT_NAME : 'BALLPACK',
        KEYOBJECT_TYPE : 'COMPOSITION'
      }
    },
    CONSTRAINTS : {
      CONSTRAINT : {
        CONSTRAINT_DEFINITION : 'HOSTS(BALLPACK)',
```

```
                CONSTRAINT_NAME : 'HOSTS'
            }
        },
        CATALOG_NAME : 'UNIVERSE'
    }
}
```

SHOW STORES

The SHOW STORES statement returns the set of stores for all catalogs in the current schema.

```
SHOW STORES [ FROM [ SCHEMA ] <schema_name> |
              FOR [ CATALOG ] <qualified_catalog_name> |
              ALL ]
```

Options:

- FROM [SCHEMA] <schema_name> - show all stores from specified schema only. The SCHEMA keyword is optional.
- FOR [CATALOG] <qualified_catalog_name> - show all stores for specified catalog. The CATALOG keyword is optional.
- ALL – show all stores from all schemas.

When the schema name or the ALL keyword are not specified, the current one is used.

Permissions

The statement requires SHOW STORE permission.

Examples

```
SHOW STORES FROM public
```

The output:

```
{
    STORE_DESCRIPTION: {
        SCHEMA_NAME: 'PUBLIC',
        CATALOG_NAME: 'UNIVERSE',
        STORE_NAME: 'BILLIARD_STORE'
    ...
},
{
    STORE_DESCRIPTION: {
        SCHEMA_NAME: 'PUBLIC',
```

```
        CATALOG_NAME: 'sales',
        STORE_NAME: 'sales_2020_Q2'
    ...
}
```

Other examples

```
SHOW STORES FOR public.universe;
SHOW STORES ALL;
```

LIST STORES

The LIST STORES statement returns the set of stores for all catalogs in the current schema.

```
LIST STORES [ FROM [ SCHEMA ] <schema_name> |
              FOR [ CATALOG ] <qualified_catalog_name> |
              ALL ]
```

Options:

- FROM [SCHEMA] <schema_name> - list all stores from specified schema only. The SCHEMA keyword is optional
- FOR [CATALOG] <qualified_catalog_name> - list all stores for specified catalog. The CATALOG keyword is optional
- ALL – list all stores from all schemas

When the schema name or the ALL keyword are not specified, the current schema is used.

Permissions

The statement requires LIST STORE permission.

Examples

```
LIST STORES FROM public
```

The output:

```
{ STORE_NAME: 'BILLIARD_STORE' },
{ STORE_NAME: 'sales_2020_Q2' },
...
```

Other examples

```
LIST STORES;
```

```
LIST STORES FOR public.universe;
LIST STORES ALL;
```

DESCRIBE STORE

The DESCRIBE STORE statement returns the KeySQL script to create the specified store.

```
DESCRIBE STORE <qualified_store_name>
```

When schema name is not specified, the current schema is used.

Permissions

The statement requires SHOW STORE permission.

Examples

```
DESCRIBE STORE public.office_rentals
```

Result:

```
{
  SCRIPT : '
-- store PUBLIC.OFFICE_RENTALS

CREATE STORE OFFICE_RENTALS FOR CATALOG UNIVERSE;'
}
```

3.8   RENAME statements

RENAME

The RENAME statement renames the specified KeySQL server object. It may be considered as a shorthand for ALTER statement with NAME option.

```
RENAME SCHEMA <old_name> TO <new_name>

RENAME (CATALOG | STORE ) <old_name> TO <new_name>
   [ IN [SCHEMA] <schema_name> ]

RENAME (KEYOBJECT | SEQUENCE) <old_name> TO <new_name>
   IN [CATALOG] [<schema_name>.]<catalog_name>

RENAME CONSTRAINT <old_name> TO <new_name>
   IN [STORE] [<schema_name>.]<store_name>
```

Options

- `<old_name>` is the name of existing KeySQL server object, i.e., the name of a schema, a constraint, etc.
- `<new_name>` is the new name for the specified object.

Permissions

The statement requires ALTER permission on the affected KeySQL object.

Examples

```
RENAME SCHEMA sales TO sales_2020;

RENAME CATALOG emps TO employees; -- in the current schema
RENAME CATALOG emps TO employees IN SCHEMA hr;
RENAME CATALOG emps TO employees IN hr;

RENAME STORE st_emps TO emps_2020 IN hr;

RENAME SEQUENCE seq_emp TO seq_emp_old IN CATALOG employees;
RENAME SEQUENCE seq_emp TO seq_emp_old IN employees;

RENAME KEYOBJECT name TO emp_name IN emps;
```

3.9 Expressions

A simple expression can be a single constant, a name of elementary k-object, or a scalar function (see KeySQL functions and operators).

An expression is a combination of simple expressions and operators.

Consider the expression grammar is as follows.

```
<expression> ::=
   <constant> | <scalar_function> |
<qualified_elementary_keyobject_name>
   | "(" <expression> ")"
   | <unary_operator> <expression>
   | <expression> <binary_operator> <expression>
```

A qualified k-object name may contain names of the corresponding schema and store. See "Query scope" and "Qualifiers" for more details.

A string expression is the expression returning the character string.

A numeric expression returns a numeric value.

Constant expressions

A constant expression is the expression that contains any valid KeySQL constants, operators, and functions only.

Examples of constant expressions:

```
'ABC'
123.45
'2022-01-02'
(123 + 45) / 67
CONCAT('ABC', 'DEF')
```

CASE expression

The CASE expression evaluates a list of conditions and returns one of the possible results. There are two forms of CASE syntax:

- The simple CASE expression compares an input expression to a set of simple WHEN-expressions to determine the result.
- The searched CASE expression evaluates a set of Boolean expressions to determine the result.

Syntax

```
-- Simple CASE expression
CASE <input_expression>
    WHEN <when_expression1> THEN <result_expression1> [ ...n ]
    [ ELSE <else_result_expression> ]
END

-- Searched CASE expression
CASE
    WHEN <boolean_expression1> THEN <result_expression1> [ ...n ]
    [ ELSE <else_result_expression> ]
END
```

Examples

Example 1: simple CASE

```
SELECT
    CASE ball.color
        WHEN 'red' THEN 'R'
        WHEN 'white' THEN 'W'
        ELSE 'X'
```

```
      END AS $color_code
FROM blrd_store
LIMIT 3;
```

Returns

```
{ COLOR_CODE : 'R' },
{ COLOR_CODE : 'W' },
{ COLOR_CODE : 'X' }
```

Example 2: searched CASE

```
SELECT
   CASE
      WHEN ball.color = 'red' AND ball.num IS NULL   THEN 1
      WHEN ball.color = 'white' OR ball.color = 'white_spot' THEN 2
      ELSE 0
   END AS $rank
FROM blrd_store
WHERE game = 'carom'
```

returns

```
{ RANK : 1 },
{ RANK : 2 },
{ RANK : 2 }
```

Example 3: CASE with constant expressions only

```
SELECT
   CASE
      WHEN EXTRACT(second FROM CURRENT_DATE()) < 20   THEN 'First'
      WHEN EXTRACT(second FROM CURRENT_DATE()) >= 40   THEN 'Third'
      ELSE 'Second'
   END AS $range
```

Returns

```
{ RANGE : 'Second' }
```

3.10   Data manipulation statements

The data manipulation statements are as follows.

INSERT INSTANCES
DELETE

UPDATE
SELECT
INSERT SELECT

INSERT INSTANCES

The statement inserts all k-object instances from the list into the store. The instances must belong to the k-objects from the catalog the store is defined for. The instances inserted into a store are called *host instances.*

```
INSERT INTO <qualified store name> INSTANCES <list of keyobject
instances>

<list of keyobject instances> ::= <keyobject instance>, > [, … ]
<keyobject instance> ::=  { <pair> }
<pair> ::= <keyobject name> : <value>
<value> ::= NULL | <constant expression> | { <members> }
<members> ::= <keyobject instance>  [, … ]
```

Permissions

The statement requires `INSERT` permissions (store scope).

Example

```
INSERT INTO MY_SCHEMA.MY_STORE INSTANCES
{LAST_NAME:'Johnson'},
{AGE:20},
{PERSON:{LAST_NAME:'Doe',FIRST_NAME:'Jane',DOB:'2001-01-01'}},
{PEOPLE:{PERSON:{LAST_NAME:'Doe',FIRST_NAME:'John',DOB:'2000-12-31'},
        PERSON:{LAST_NAME:'Doe',FIRST_NAME:'Jane',DOB:'2001-01-
01'}}};
```

Instance ID and Version

When inserted into a store, each host instance is complemented with two default system attributes, which are the instances of the reserved k-objects named `_IID`, and `_VERSION` that are not present in any catalog.

The `_IID` represents a unique (within the store) and permanent instance identifier. The `_VERSION` is an incremental version number of the instance – any update of an instance increments its version by 1.

The values of both host instance attributes can be retrieved like the values of user-defined k-objects with some limitations. Refer to SELECT statement for more details.

DELETE

The DELETE statement deletes each host instance in the store subject to the optional scope and WHERE clause condition.

```
DELETE FROM <qualified store name> [(scope)]
[ WHERE <condition> ]
```

If the scope is present, only the host instances which are the instances of the scope k-object are deleted subject to WHERE clause. The "out of scope" host instances are not deleted.

If the WHERE clause is missing, all instances are deleted from the store subject to the scope if any.

WHERE clause is evaluated only on host instances that reference all k-objects the statement references; the statement has no effect on all other host instances.

Refer to SELECT statement for more details on the WHERE clause semantics.

Permissions

The statement requires DELETE permissions (store scope).

Examples

Consider my_store containing four host instances formed by the INSERT example above.

```
TRUNCATE my_store; -- clear the store
INSERT INTO my_store INSTANCES
{last_name: 'Johnson'},
{age: 20},
{person: {last_name:'Doe', first_name: 'Jane', DOB: '2001-01-01'}},
{people: {person: {last_name: 'Doe', first_name: 'John', DOB: '2000-
12-31'},
        person: {last_name: 'Doe', first_name: 'Jane', DOB: '2001-
01-01'}}};
```

Example 1

Consider the following DELETE statement.

```
DELETE FROM my_store(person)
-- DELETE query OK, 1 instance affected
```

This means that only the host instance of the `person` k-object should be deleted.

To see the resulting state of `my_store`:

```
SELECT * FROM my_store
```

The result will be as follows.

```
{
  LAST_NAME : 'Johnson'
},
{
  AGE : 20
},
{
  PEOPLE : {
    PERSON : {
      LAST_NAME : 'Doe',
      FIRST_NAME : 'John',
      DOB : '2000-12-31 00:00:00'
    },
    PERSON : {
      LAST_NAME : 'Doe',
      FIRST_NAME : 'Jane',
      DOB : '2001-01-01 00:00:00'
    }
  }
}
```

Example 2

Restore data using `INSERT` statement above.

Run the following `DELETE` statement:

```
DELETE FROM my_store
WHERE first_name = 'Jane'
-- DELETE query OK, 2 instances affected
```

All host instances containing `first_name` equal to `'Jane'` will be deleted.

```
SELECT * FROM my_store
```

The resulting store is as follows.

```
{
  LAST_NAME : 'Johnson'
```

```
},
{
  AGE : 20
}
```

DELETE by Instance ID and Version

WHERE clause condition can also reference host instance attributes _IID and
_VERSION along with regular user-defined k-objects.

Restore data and run the following query:

```
DELETE FROM my_store
WHERE last_name IS NULL
      AND _IID >= 9 AND _VERSION >= 1
-- DELETE query OK, 0 instances affected
```

UPDATE

The statement updates each host instance in the store subject to the optional
scope and WHERE clause condition.

```
UPDATE <qualified store name> [ ( <scope> ) ]
SET <list of assignments>
[ WHERE <condition> ]

<scope> ::= <keyobject name>
<list of assignments> ::= <assignment> [, <list of assignments> ]
<assignment>::=
    <elementary or composition keyobject assignment> |
    <multi-composition keyobject assignment>
<elementary or composition keyobject assignment>::=
    <keyobject name> = <simple expression>
<multi-composition keyobject assignment>::=
    <keyobject name> += <value> |
    <keyobject name> -= <value>
```

If a host instance does not reference a k-object specified by the scope, the
host instance is not updated. If a host instance does reference the scope k-
object, all updates are made exclusively within the instances of the scope k-
object contained by the host instance.

Only the host instances that reference all k-objects specified in the SET clause
are considered for the update; the statement has no effect on other host
instances.

WHERE clause is formed and applied according to the specification of SELECT statement. WHERE clause is evaluated only on host instances that reference all k-objects the WHERE clause references; the statement has no effect on other host instances.

Refer to SELECT statement for more details.

Permissions

The statement requires UPDATE permissions (store scope).

Examples

Consider my_catalog defining following k-objects:

```
CREATE CATALOG my_catalog;
CREATE KEYOBJECT age NUMBER IN CATALOG my_catalog;
CREATE KEYOBJECT first_name CHAR IN CATALOG my_catalog;
CREATE KEYOBJECT last_name CHAR IN CATALOG my_catalog;
CREATE KEYOBJECT DOB DATE IN CATALOG my_catalog;
CREATE KEYOBJECT person {last_name, first_name, DOB} IN CATALOG
my_catalog;
CREATE KEYOBJECT people {person MULTIPLE} IN CATALOG my_catalog;
```

The store my_store is also defined for the catalog:

```
CREATE STORE my_store FOR CATALOG my_catalog;
```

Consider MY_STORE containing four host instances formed by the following INSERT statement.

▌ This code will also be used to reinitialize the data for the examples below.

```
TRUNCATE my_store; -- clear the store
INSERT INTO my_store INSTANCES
{last_name: 'Johnson'},
{age: 20},
{person: {last_name:'Doe', first_name: 'Jane', DOB: '2001-01-01'}},
{people:
{person: {last_name: 'Doe', first_name: 'John', DOB: '2000-12-31'},
 person: {last_name: 'Doe', first_name: 'Jane', DOB: '2001-01-01'}}};
```

Example 1

Consider the following UPDATE statement.

```
UPDATE my_store
SET last_name = NULL
```

```
-- UPDATE query OK, 3 instances affected
```

To see the resulting state of my_store run

```
SELECT * FROM my_store
```

The result will be as follows.

```
{
  LAST_NAME : NULL
},
{
  AGE : 20
},
{
  PERSON : {
    LAST_NAME : NULL,
    FIRST_NAME : 'Jane',
    DOB : '2001-01-01 00:00:00'
  }
},
{
  PEOPLE : {
    PERSON : {
      LAST_NAME : NULL,
      FIRST_NAME : 'John',
      DOB : '2000-12-31 00:00:00'
    },
    PERSON : {
      LAST_NAME : NULL,
      FIRST_NAME : 'Jane',
      DOB : '2001-01-01 00:00:00'
    }
  }
}}
```

All host instances containing k-object last_name have been updated with the value NULL.

Example 2

Reinitialize the store as mentioned above.

Consider the following UPDATE statement with a scope.

```
UPDATE my_store(people)
SET LAST_NAME = NULL
-- UPDATE query OK, 1 instance affected
```

Now the resulting state of my_store is as follows.

```
{
   LAST_NAME : 'Johnson'
},
{
   AGE : 20
},
{
   PERSON : {
      LAST_NAME : 'Doe',
      FIRST_NAME : 'Jane',
      DOB : '2001-01-01 00:00:00'
   }
},
{
   PEOPLE : {
      PERSON : {
         LAST_NAME : NULL,
         FIRST_NAME : 'John',
         DOB : '2000-12-31 00:00:00'
      },
      PERSON : {
         LAST_NAME : NULL,
         FIRST_NAME : 'Jane',
         DOB : '2001-01-01 00:00:00'
      }
   }
}
```

As we can see, the updates are only performed within the instance of the k-object people which happens to be a host instance.

Example 3

Reinitialize the store. Now we are changing the scope as follows.

```
UPDATE my_store(person)
SET last_name = last_name || '_updated'
-- UPDATE query OK, 1 instance affected.
```

The statement modifies all instances of the k-object person within the host instances as follows.

```
{
   LAST_NAME : 'Johnson'
```

```
  },
  {
    AGE : 20
  },
  {
    PERSON : {
      LAST_NAME : 'Doe_updated',
      FIRST_NAME : 'Jane',
      DOB : '2001-01-01 00:00:00'
    }
  },
  {
    PEOPLE : {
      PERSON : {
        LAST_NAME : 'Doe',
        FIRST_NAME : 'John',
        DOB : '2000-12-31 00:00:00'
      },
      PERSON : {
        LAST_NAME : 'Doe',
        FIRST_NAME : 'Jane',
        DOB : '2001-01-01 00:00:00'
      }
    }
  }
}
```

Example 4

Reinitialize the store. Now we are adding or removing a value to/from the instance of a multi-composition k-object.

```
UPDATE my_store
SET people += {person: {last_name: 'New', first_name: 'New', DOB:
NULL}}
-- UPDATE query OK, 1 instance affected
```

The statement appends the specified person instance to all people instances (only one is in our example) as follows.

```
{
  LAST_NAME : 'Johnson'
},
{
  AGE : 20
},
{
  PERSON : {
    LAST_NAME : 'Doe',
```

```
        FIRST_NAME : 'Jane',
        DOB : '2001-01-01 00:00:00'
      }
    },
    {
      PEOPLE : {
        PERSON : {
          LAST_NAME : 'Doe',
          FIRST_NAME : 'John',
          DOB : '2000-12-31 00:00:00'
        },
        PERSON : {
          LAST_NAME : 'Doe',
          FIRST_NAME : 'Jane',
          DOB : '2001-01-01 00:00:00'
        },
        PERSON : {
          LAST_NAME : 'New',
          FIRST_NAME : 'New',
          DOB : NULL
        }
      }
    }
```

To remove an instance, from all people instances where it is present:

```
UPDATE my_store
SET people -= {person: {last_name: 'New', first_name: 'New', DOB:
NULL}}
-- UPDATE query OK, 1 instance affected
```

Result will return the initial store.

```
{
  LAST_NAME : 'Johnson'
},
{
  AGE : 20
},
{
  PERSON : {
    LAST_NAME : 'Doe',
    FIRST_NAME : 'Jane',
    DOB : '2001-01-01 00:00:00'
  }
},
{
  PEOPLE : {
    PERSON : {
```

```
      LAST_NAME : 'Doe',
      FIRST_NAME : 'John',
      DOB : '2000-12-31 00:00:00'
    },
    PERSON : {
      LAST_NAME : 'Doe',
      FIRST_NAME : 'Jane',
      DOB : '2001-01-01 00:00:00'
    }
  }
}
```

Example 5

Reinitialize the store.

Several update actions can be performed at once provided that all the actions can be applied to the same instance. The actions are performed in the order they are listed in the query as follows.

```
UPDATE my_store
SET DOB = '2000-12-31',
    people += {person: {last_name: 'New', FIRST_NAME: 'New', DOB:
NULL}}
-- UPDATE query OK, 1 instance affected
```

The resulting store is as follows.

```
{
  LAST_NAME : 'Johnson'
},
{
  AGE : 20
},
{
  PERSON : {
    LAST_NAME : 'Doe',
    FIRST_NAME : 'Jane',
    DOB : '2001-01-01 00:00:00'
  }
},
{
  PEOPLE : {
    PERSON : {
      LAST_NAME : 'Doe',
      FIRST_NAME : 'John',
      DOB : '2000-12-31 00:00:00'
    },
```

```
    PERSON : {
      LAST_NAME : 'Doe',
      FIRST_NAME : 'Jane',
      DOB : '2000-12-31 00:00:00'
    },
    PERSON : {
      LAST_NAME : 'New',
      FIRST_NAME : 'New',
      DOB : NULL
    }
  }
}
```

Please note that the DOB of the third host instance (person) was not updated because it does not belong to an instance of people. Only the host instances that reference all k-objects specified in the SET clause are considered for the update; the statement has no effect on all other host instances.

UPDATE by Instance ID and Version

WHERE clause condition can also reference host instance attributes _IID and _VERSION along with regular user-defined k-objects.

Example

```
UPDATE my_store
SET DOB = NULL
WHERE last_name IS NOT NULL AND _IID >= 9 AND _VERSION >= 1
-- UPDATE query OK, 2 instances affected
```

The resulting store is

```
{
  LAST_NAME : 'Johnson'
},
{
  AGE : 20
},
{
  PERSON : {
    LAST_NAME : 'Doe',
    FIRST_NAME : 'Jane',
    DOB : NULL
  }
},
{
  PEOPLE : {
```

```
      PERSON : {
        LAST_NAME : 'Doe',
        FIRST_NAME : 'John',
        DOB : NULL
      },
      PERSON : {
        LAST_NAME : 'Doe',
        FIRST_NAME : 'Jane',
        DOB : NULL
      }
    }
  }
}
```

SELECT

The SELECT statement returns a set of k-object instances (a store) produced from the host instances inserted into one or more stores.

The UNION, INTERSECT, and MINUS operators can be used between the SELECT queries to combine their outputs into one result set.

```
SELECT <select_list>
FROM <from_sources>
[ WHERE <search_condition> ]
[ GROUP BY <keyobject_expression_list> ]
[ HAVING <search_condition> ]
[ ORDER BY <order_by_list> ]
[ LIMIT <output_limit> ]

<select_list> ::= <output_list> | [ DISTINCT ] "*" | <expression>
<output_list> ::=
    <expression>
    | [ DISTINCT ] <qualified_keyobject_name>
    | "{" <output_list> "}"
        [ ( AS | TO ) ( <keyobject_name> | <adhoc_name> ) ] [ ","
<output_list> ]

<qualified_keyobject_name> ::=
    [ ( <store_source> | <source_alias> ) ( "." | ".." ) ]
<keyobject_path>
<keyobject_path> ::=
    <keyobject_name> [ "[" <instance_marker> "]" ] [ "."
<keyobject_path> ]
<instance_marker> ::= <simple_identifier>

<adhoc_name> ::= "$" <simple_identifier>

<keyobject_name_list> ::=
    <qualified_keyobject_name> [ "," <keyobject_name_list> ]
```

```
<source_alias> ::= <identifier>
<value_expression> ::=
    <constant>
    | <scalar_function> "(" [ <function_args> ] ")"

<from_sources> ::= <store_source> [ <joined_store_sources> ]
<store_source> ::= <qualified_store_name> [ "(" <scope> ")" ] [
<source_alias> ]
<joined_store_sources> ::=
    <source_join> <store_source> ON <join_conditions> [
<joined_store_sources> ]
<source_join> ::=
    JOIN | LEFT JOIN

<search_condition> ::=
    [ NOT ] "(" <predicate> ")" | <search_condition>
    [ ( AND | OR ) <predicate> | <search_condition> ]
<predicate> ::=
    <expression> "=" | "<>" | ">" | ">=" | "<" | "<=" <expression>
    | <string_expression> LIKE <string_expression>
    | NOT "(" <string_expression> LIKE <string_expression> ")"
    | <expression> IS [ NOT ] NULL
    | <expression> [ NOT ] IN "(" <value_list> ")"

<value_list> ::= <constant_expression> [ "," <value_list> ]

<order_by_list> ::=
    <expression> [ ASC | DESC ] [ "," <order_by_list> ]

<output_limit> ::= 1 .. <max_int64_value>
```

Arguments:

- AS <keyobject_name> allows to restructure instances in a resulting store.
  See "Restructuring projection results: operator AS" and "Using operator
  AS for flattening" for more details and examples.
- TO <keyobject_name> allows to rename instances in a resulting store. See
  "Renaming projection results" for more details and examples.
- AS | TO <adhoc_name> allows to output the query result as temporary k-
  objects without creating them in the catalog. See "User-defined ad hoc k-
  objects" for more details and examples.
- SELECT <expression> calculates the expression result as a single k-object
  instance. The result then may be cast with AS to an existing or an ad hoc k-
  object. See "Expressions" for more details.

> Note that SELECT <constant_expression> does not allow the FROM clause.

See "SELECT without FROM" for more details and examples.

- JOIN and LEFT JOIN are SQL-like joins which semantics is different from the SQL ones.
- An expression specifying the <qualified_keyobject_name> can also contain two dots qualifier which means "*one or more nesting levels*". See example 3 "Specify the nesting level of a keyobject in the query".
- The <keyobject_path> with <instance_marker> are described above in "Instance markers".
- The modifier DISTINCT can be applied to any number of k-objects in the <select_list>. If it is not applied to each k-object in the list, the rest of the k-objects are subject to GROUP BY. See "Operator GROUP BY " and "Operator DISTINCT" for more details and examples.
- The GROUP BY clause divides the result into groups of k-objects for the purpose of performing one or more aggregations on each group. See "Operator GROUP BY " and "Aggregate functions".
- The HAVING clause is typically used with a GROUP BY one. When GROUP BY is not used, there is an implicit single, aggregated group. Search conditions of HAVING clause may contain aggregate functions (COUNT, SUM etc.) whereas the WHERE clause cannot.
- The IN predicate may contain scalar values of different types, i.e., WHERE color IN ('red', 16711680)

The order in which k-object instances are returned in a result set is not guaranteed unless the ORDER BY clause is present.

> Subqueries will be available in the future versions of KeySQL.

Permissions

The statement requires SELECT permissions (store scope).

Example 1. Querying a store

Consider the catalog `people`:

```
CREATE CATALOG people;

CREATE KEYOBJECT city CHAR IN CATALOG people;
CREATE KEYOBJECT country CHAR IN CATALOG people;
CREATE KEYOBJECT email_address CHAR IN CATALOG people;
```

```
CREATE KEYOBJECT email_type CHAR IN CATALOG people;
CREATE KEYOBJECT first_name CHAR IN CATALOG people;
CREATE KEYOBJECT last_name CHAR IN CATALOG people;
CREATE KEYOBJECT YOB NUMBER IN CATALOG people;
CREATE KEYOBJECT cnt NUMBER IN CATALOG people;
CREATE KEYOBJECT address {city, country} IN CATALOG people;
CREATE KEYOBJECT email {email_address, email_type} IN CATALOG people;
CREATE KEYOBJECT emails {email MULTIPLE} IN CATALOG people;
CREATE KEYOBJECT person {address, emails, first_name, last_name, YOB}
IN CATALOG people;
```

Consider the store `people_store` containing the data.

```
CREATE STORE people_store FOR CATALOG people;
INSERT INTO people_store INSTANCES
{
    person: {
        emails: {
            email: {email_address: 'shopping@johns.com', email_TYPE:
NULL},
            email: {email_address: 'amazon.shopping@johns.com',
email_TYPE: NULL}
        },
        first_name: NULL,
        last_name: 'Johns',
        YOB: 1999,
        address: {country: NULL, city: NULL}
    }
},
{
    person: {
        emails:{
            email: {email_address: 'hobby@miller.com', email_type:
NULL}
        },
        first_name: 'John',
        last_name:'Miller',
        YOB: 1999,
        address: {country: NULL, city:NULL}
    }
},
{
    person: {
        emails: {
            email: {email_address: 'smith@work.org', email_type:
'work'},
            email: {email_address: 'John@smith.org', email_type:
'home'}
        },
```

```
        first_name: 'John',
        last_name: 'Smith',
        YOB: 1997,
        address: {country: 'US', city:'San-Francisco'}
    }
};
```

The query:

```
SELECT * FROM people_store
```

Produces all host instances from the `people_store`.

The query:

```
SELECT PERSON FROM people_store
```

is equivalent to the previous query because people_store contains only host instances of type PERSON.

The result of the query:

```
SELECT first_name, last_name FROM people_store
```

will be:

```
{RESULT:{FIRST_NAME:NULL,LAST_NAME:'Johns'}},
{RESULT:{FIRST_NAME:'John',LAST_NAME:'Miller'}},
{RESULT:{FIRST_NAME:'John',LAST_NAME:'Smith'}}
```

The result of the query:

```
SELECT last_name, email_address
FROM people_store
WHERE first_name = 'John';
```

will be:

```
{RESULT:{LAST_NAME:'Miller',EMAIL_ADDRESS:'hobby@miller.com'}},
{RESULT:{LAST_NAME:'Smith',#EMAIL_ADDRESS:{EMAIL_ADDRESS:'smith@work.
org',

EMAIL_ADDRESS:'John@smith.org'}}}
```

The result of the query:

```
SELECT email_address, email_type
```

```
FROM people_store
WHERE first_name = 'John'
GROUP BY email_address
ORDER BY email_type
```

will be:

```
{RESULT:{EMAIL_ADDRESS:'hobby@miller.com',EMAIL_TYPE:NULL}},
{RESULT:{EMAIL_ADDRESS:'John@smith.org',EMAIL_TYPE:'home'}},
{RESULT:{EMAIL_ADDRESS:'smith@work.org',EMAIL_TYPE:'work'}}
```

Note that the set of k-objects in GROUP BY (email_address) should be present in the SELECT list.

The set of k-objects in ORDER BY (email_address) should satisfy several conditions:

- Should be present in SELECT list or be a member of any member of the list.
- Should be an elementary k-object:
  - ORDER BY email_type is OK.
  - ORDER BY email is not OK.

To illustrate this, consider a query where email_type is not present in SELECT list but is the member of email.

```
SELECT email, email_address
FROM people_store
WHERE first_name = 'John'
GROUP BY email_address
ORDER BY email_type
```

The result of this query will be:

```
{RESULT:{EMAIL:{EMAIL_ADDRESS:'hobby@miller.com',EMAIL_TYPE:NULL},
        EMAIL_ADDRESS:'hobby@miller.com'}},
{RESULT:{EMAIL:{EMAIL_ADDRESS:'John@smith.org',EMAIL_TYPE:'home'},
        EMAIL_ADDRESS:'John@smith.org'}},
{RESULT:{EMAIL:{EMAIL_ADDRESS:'smith@work.org',EMAIL_TYPE:'work'},
        EMAIL_ADDRESS:'smith@work.org'}}
```

The ORDER BY [ASC/DESC] (ascending/descending) is used to sort the output list.

The result of the query:

```
SELECT last_name, count(email)
```

```
FROM people_store
GROUP BY last_name;
```

will be:

```
{RESULT:{LAST_NAME:'Smith',COUNT_EMAIL:2}},
{RESULT:{LAST_NAME:'Johns',COUNT_EMAIL:2}},
{RESULT:{LAST_NAME:'Miller',COUNT_EMAIL:1}}
```

The result of the query:

```
SELECT last_name, count(email)
FROM people_store
GROUP BY last_name
ORDER BY last_name ASC
```

will be:

```
{RESULT:{LAST_NAME:'Johns',COUNT_EMAIL:2}},
{RESULT:{LAST_NAME:'Miller',COUNT_EMAIL:1}},
{RESULT:{LAST_NAME:'Smith',COUNT_EMAIL:2}}
```

Finally, the result of the query:

```
SELECT last_name, count(email)
FROM people_store
GROUP BY last_name
ORDER BY last_name DESC
```

will be:

```
{RESULT:{LAST_NAME:'Smith',COUNT_EMAIL:2}},
{RESULT:{LAST_NAME:'Miller',COUNT_EMAIL:1}},
{RESULT:{LAST_NAME:'Johns',COUNT_EMAIL:2}}
```

If we modify this query as follows:

```
SELECT last_name, count(email)
FROM people_store
GROUP BY last_name
ORDER BY last_name DESC
LIMIT 2
```

the output will be limited to 2 instances:

```
{RESULT:{LAST_NAME:'Miller',COUNT_EMAIL:1}},
{RESULT:{LAST_NAME:'Johns',COUNT_EMAIL:2}}
```

To select only the resulting records satisfying specific conditions HAVING can be used.

```
SELECT last_name, count(email) as cnt
FROM people_store
GROUP BY last_name
HAVING cnt > 1
ORDER BY last_name ASC
```

The result will be:

```
{RESULT:{LAST_NAME:'Johns',CNT:2}},
{RESULT:{LAST_NAME:'Smith',CNT:2}}
```

Example 2: DISTINCT and GROUP BY

```
SELECT DISTINCT first_name
FROM public.people_store
;
SELECT DISTINCT last_name, DISTINCT first_name
FROM public.people_store
;
SELECT DISTINCT last_name, first_name
FROM public.people_store
GROUP BY first_name;
```

Those statements produce the following respective results.

```
{
    FIRST_NAME : 'John'
},
{
    FIRST_NAME : NULL
}
-----------------------------------
{
    RESULT : {
        #LAST_NAME : {
            LAST_NAME : 'Smith',
            LAST_NAME : 'Johns',
            LAST_NAME : 'Miller'
        },
        #FIRST_NAME : {
            FIRST_NAME : 'John',
            FIRST_NAME : NULL
        }
    }
}
```

```
-----------------------------------
{
  RESULT : {
    #LAST_NAME : {
      LAST_NAME : 'Smith',
      LAST_NAME : 'Miller'
    },
    FIRST_NAME : 'John'
  }
},
{
  RESULT : {
    LAST_NAME : 'Johns',
    FIRST_NAME : NULL
  }
}
}
```

Example 3. Specifying the nesting level of a keyobject in the query

Consider the catalog shopping containing the names of shops and products on different nesting levels.

```
CREATE CATALOG shopping;
CREATE KEYOBJECT sku CHAR IN CATALOG shopping;
CREATE KEYOBJECT name CHAR IN CATALOG shopping;
CREATE KEYOBJECT product {sku, name} IN CATALOG shopping;
CREATE KEYOBJECT products {product MULTIPLE} IN CATALOG shopping;
CREATE KEYOBJECT shop {name, products} IN CATALOG shopping;
```

Consider the store shops containing the following data.

```
CREATE STORE shops FOR CATALOG shopping;
INSERT INTO shops INSTANCES
{
  shop: {
  name: 'Croissant',
    products: {
      product: {sku: 'CR01', name: 'Apple pie'},
      product: {sku: 'CR02', name: 'Donuts'},
      product: {sku: 'DN02', name: 'Cookies'}
    }
  }
},
{
  shop: {
  name: 'Apple pie',
    products: {
```

```
        product: {sku: 'AP01', name: 'Croissant'},
        product: {sku: 'AP02', name: 'Donuts'},
        product: {sku: 'AP03', name: 'Apple pie'}
      }
    }
  },
  {
    shop: {
    name: 'Techno',
      products: {
        product: {sku: 'TC01', name: 'DIY kit'}
    }
    }
  },
  {
    shop: {
    name: 'Donuts',
      products: {
        product: {sku: 'DN01', name: 'Donuts'},
        product: {sku: 'DN02', name: 'Apple pie'},
        product: {sku: 'DN02', name: 'Brownies'}
      }
    }
  }
};
```

Q1. Select all distinct names of both shops and products:

```
SELECT DISTINCT name
FROM shops
ORDER BY name
```

The query returns:

```
{ NAME : 'Apple pie' },
{ NAME : 'Brownies' },
{ NAME : 'Cookies' },
{ NAME : 'Croissant' },
{ NAME : 'DIY kit' },
{ NAME : 'Donuts' },
{ NAME : 'Techno' }
```

Q2. Same as Q1 but using two dots to explicitly specify the top scope of names:

```
SELECT DISTINCT shop..name
FROM shops
ORDER BY name
```

Returns the same result as Q1 does.

Q3. Select only shop names:

```
SELECT DISTINCT shop.name
FROM shops
ORDER BY name
```

The query returns:

```
{ NAME : 'Apple pie' }
{ NAME : 'Croissant' },
{ NAME : 'Donuts' },
{ NAME : 'Techno' },
```

Q4. Select only product names:

```
SELECT DISTINCT product.name
FROM shops
ORDER BY name
```

The query returns:

```
{ NAME : 'Apple pie' },
{ NAME : 'Brownies' },
{ NAME : 'Cookies' },
{ NAME : 'Croissant' },
{ NAME : 'DIY kit' },
{ NAME : 'Donuts' }
```

Q5. Select shops having the specified name or selling the products with the specified name.

First attempt is wrong:

```
SELECT shop
FROM shops
WHERE name = 'Apple pie'
```

 Error: ambiguous reference to NAME

Explicit specification returns as expected.

```
SELECT shop.name -- to reduce the output we only ask for the shop
names
FROM shops
WHERE shop.name = 'Apple pie' OR  product.name = 'Apple pie'
```

The result is:

```
{
  NAME : 'Croissant'
},
{
  NAME : 'Apple pie'
},
{
  NAME : 'Donuts'
}
```

Example 4. Working with supertype/subtype structures

A deeper dive example created by Andris Korzans. The reader can skip it in the first reading.

Modeling rules:

*   The attributes in the supertype are common to all subtypes.
*   The attributes in the subtype apply only to that subtype.

In this example the supertype CUSTOMER has two subtypes: ORGANIZATION and INDIVIDUAL.

```
-- Create catalog
CREATE CATALOG X;
-- Create elementary k-objects
CREATE KEYOBJECT GUID CHAR IN CATALOG X;
CREATE KEYOBJECT NAME CHAR IN CATALOG X;
CREATE KEYOBJECT FIRST_NAME CHAR IN CATALOG X;
CREATE KEYOBJECT MIDDLE_NAME CHAR IN CATALOG X;
CREATE KEYOBJECT LAST_NAME CHAR IN CATALOG X;
CREATE KEYOBJECT SUBTYPE CHAR IN CATALOG X;
CREATE KEYOBJECT TYPE CHAR IN CATALOG X;
CREATE KEYOBJECT ADDRESS CHAR IN CATALOG X;
CREATE KEYOBJECT GENDER CHAR IN CATALOG X;
CREATE KEYOBJECT BIRTH_DATE DATE IN CATALOG X;
CREATE KEYOBJECT DEATH_DATE DATE IN CATALOG X;
CREATE KEYOBJECT ESTABILISHED_DATE DATE IN CATALOG X;
CREATE KEYOBJECT CLOSE_DATE DATE IN CATALOG X;
CREATE KEYOBJECT RN CHAR IN CATALOG X;
CREATE KEYOBJECT EIN CHAR IN CATALOG X;
CREATE KEYOBJECT SSN CHAR IN CATALOG X;
CREATE KEYOBJECT ITIN CHAR IN CATALOG X;
-- Create composite k-objects
CREATE KEYOBJECT CUSTOMER {GUID, SUBTYPE, ADDRESS, EMAIL, PHONE} IN
CATALOG X;
CREATE KEYOBJECT ORGANIZATION {GUID, NAME, TYPE, ESTABLISHED_DATE,
CLOSE_DATE, RN, EIN} IN CATALOG X;
```

```
CREATE KEYOBJECT INDIVIDUAL {GUID, FIRST_NAME, MIDDLE_NAME,
LAST_NAME, BIRTH_DATE, DEATH_DATE, GENDER, SSN, ITIN} IN CATALOG X;
-- Create store
CREATE STORE XSTORE FOR CATALOG X;
-- Insert sample data
INSERT INTO XSTORE INSTANCES
 {CUSTOMER:{GUID:'cust1', SUBTYPE: 'Organization'}}
,{CUSTOMER:{GUID:'cust2', SUBTYPE: 'Individual'}}
,{ORGANIZATION:{GUID:'cust1', NAME: 'Sunrise', TYPE: 'LLC',
ESTABLISHED_DATE: '1998-01-01 00:00:00', RN: '', EIN: '00-0000000'}}
,{INDIVIDUAL: {GUID: 'cust2', FIRST_NAME: 'John', MIDDLE_NAME: 'J',
LAST_NAME: 'Johnson', BIRTH_DATE: '1992-11-03 00:00:00', GENDER: 'M',
SSN: 'NNN-NN-NNN', ITIN: '9NN-NN-NNNN'}}
;
```

There are two ways to select supertype/subtype data in the KeySQL.

First way is to join supertype k-object to all subtype k-objects using multiple
LEFT JOINS as follows:

```
SELECT c.CUSTOMER, o.ORGANIZATION, i.INDIVIDUAL
FROM XSTORE(CUSTOMER) c
LEFT JOIN XSTORE(ORGANIZATION) o ON o.GUID = c.GUID
LEFT JOIN XSTORE(INDIVIDUAL) i ON i.GUID = c.GUID
```

The result is the same:

```
{
  RESULT : {
    CUSTOMER : {
      GUID : 'cust1',
      SUBTYPE : 'Organization',
      ADDRESS : NULL,
      EMAIL : NULL,
      PHONE : NULL
    },
    ORGANIZATION : {
      GUID : 'cust1',
      NAME : 'Sunrise',
      TYPE : 'LLC',
      ESTABILISHED_DATE : '1998-01-01 00:00:00',
      CLOSE_DATE : NULL,
      RN : '',
      EIN : '00-0000000'
    },
    INDIVIDUAL : NULL
  }
},
{
```

```
  RESULT : {
    CUSTOMER : {
      GUID : 'cust2',
      SUBTYPE : 'Individual',
      ADDRESS : NULL,
      EMAIL : NULL,
      PHONE : NULL
    },
    ORGANIZATION : NULL,
    INDIVIDUAL : {
      GUID : 'cust2',
      FIRST_NAME : 'John',
      MIDDLE_NAME : 'J',
      LAST_NAME : 'Johnson',
      GENDER : 'M',
      BIRTH_DATE : '1992-11-03 00:00:00',
      DEATH_DATE : NULL,
      SSN : 'NNN-NN-NNN',
      ITIN : '9NN-NN-NNNN'
    }
  }
}
```

> The drawback of this approach: user needs to rewrite all SELECT statements when the supertype structure changes, i.e., a new subtype is added.

The second way:

- create an interfacing k-object CUSTOMER_SUBTYPE_DETAILS as the composition of all subtypes of the CUSTOMER supertype:
- join a supertype k-object to all subtype k-objects using a single join * AS CUSTOMER_SUBTYPE_DETAILS

```
CREATE KEYOBJECT CUSTOMER_SUBTYPE_DETAILS {ORGANIZATION, INDIVIDUAL}
IN CATALOG X;

SELECT c.CUSTOMER, d.* AS CUSTOMER_SUBTYPE_DETAILS
FROM XSTORE(CUSTOMER) c
JOIN XSTORE d ON c.GUID = d.GUID
```

The result is:

```
{
  RESULT : {
    CUSTOMER : {
      GUID : 'cust1',
      SUBTYPE : 'Organization',
```

```
        ADDRESS : NULL,
        EMAIL : NULL,
        PHONE : NULL
      },
      CUSTOMER_SUBTYPE_DETAILS : {
        ORGANIZATION : {
          GUID : 'cust1',
          NAME : 'Sunrise',
          TYPE : 'LLC',
          ESTABILISHED_DATE : '1998-01-01 00:00:00',
          CLOSE_DATE : NULL,
          RN : '',
          EIN : '00-0000000'
        },
        INDIVIDUAL : NULL
      }
    }
  },
  {
    RESULT : {
      CUSTOMER : {
        GUID : 'cust2',
        SUBTYPE : 'Individual',
        ADDRESS : NULL,
        EMAIL : NULL,
        PHONE : NULL
      },
      CUSTOMER_SUBTYPE_DETAILS : {
        ORGANIZATION : NULL,
        INDIVIDUAL : {
          GUID : 'cust2',
          FIRST_NAME : 'John',
          MIDDLE_NAME : 'J',
          LAST_NAME : 'Johnson',
          GENDER : 'M',
          BIRTH_DATE : '1992-11-03 00:00:00',
          DEATH_DATE : NULL,
          SSN : 'NNN-NN-NNN',
          ITIN : '9NN-NN-NNNN'
        }
      }
    }
  }
}
```

Example 5. Using _IID and _VERSION

WHERE clause condition can also reference host instance system attributes _IID and _VERSION along with the regular catalog k-objects.

Consider my_store used in the UPDATE examples. The following query returns _IID and _VERSION of all instances in the store:

```
SELECT _IID, _VERSION FROM my_store
-- 4 instances in result set
```

Result:

```
{
  RESULT : { _IID : 53, _VERSION : 0 }
},
{
  RESULT : { _IID : 54, _VERSION : 0 }
},
{
  RESULT : { _IID : 55, _VERSION : 0 }
},
{
  RESULT : { _IID : 56, _VERSION : 0 }
}
```

Then the query

```
SELECT * FROM my_store
WHERE _IID = 56
```

returns one instance of `people` k-object as expected:

```
{
  PEOPLE : {
    PERSON : {
      LAST_NAME : 'Doe',
      FIRST_NAME : 'John',
      DOB : '2000-12-31 00:00:00'
    },
    PERSON : {
      LAST_NAME : 'Doe',
      FIRST_NAME : 'Jane',
      DOB : '2001-01-01 00:00:00'
    }
  }
}
```

INSERT SELECT

The statement inserts all k-object instances produced by the SELECT query into the store. All resulting instances must belong to the k-objects from the same catalog the store is defined for.

```
INSERT INTO <qualified_store_name>
<SELECT statement>
```

Permissions

The statement requires both INSERT and SELECT permissions (store scope).

Example

```
CREATE STORE orders_2020_01 FOR CATALOG sales.orders;
INSERT INTO sales.orders_2020_01
SELECT order
FROM sales.orders_2020
WHERE order.date >= '2020-01-01' AND order.date < '2020-02-01';
```

3.11   Set operators usage and order of precedence

Set operators may be combined in an expression. The order of precedence of set operators is as follows.

- Expression inside parentheses (highest)
- INTERSECT
- UNION [ALL], MINUS

Syntax

```
<query_specification> | "(" <query_expression> ")"
( UNION [ ALL ] | INTERSECT | MINUS )
<query_specification> | "(" <query_expression> ")"
[ ORDER BY <order_by_list> ]
[ LIMIT <output_limit> ]
```

`<query_specification> | "(" <query_expression> ")"` is a SELECT query or a query expression that returns the data to be combined with the data from another query or expression.

The result set can be ordered and limited using ORDER BY and LIMIT clauses.

Examples

```
-- Example 1
SELECT game FROM blrd_store WHERE ball.color = 'red'
UNION
SELECT game FROM blrd_store WHERE ball.color = 'blue'
--
-- Example 2
SELECT game FROM blrd_store WHERE ball.color = 'red'
INTERSECT
SELECT game FROM blrd_store WHERE ball.color = 'blue'
--
-- Example 3
SELECT game FROM blrd_store WHERE ball.color = 'red'
MINUS
SELECT game FROM blrd_store WHERE ball.color = 'blue'
--
-- Example 4
(
  (
    (
      SELECT game FROM blrd_store WHERE ball.color = 'red'
      UNION
      SELECT game FROM blrd_store WHERE ball.color = 'blue'
    )
    INTERSECT
    SELECT game FROM blrd_store WHERE ball.color = 'blue'
  )
  MINUS
  SELECT game FROM blrd_store WHERE ball.color = 'ivory'
)
ORDER BY game
LIMIT 100
```

UNION

The UNION operator combines the results of two SELECT statements into a single result set. The user can control whether the result set includes or eliminates duplicate host k-object instances:

UNION eliminates duplicate instances.
UNION ALL includes duplicate instances.

Both operations are commutative – the result does not depend on the order of the operands.

Example 1

```
SELECT 1 as $Y
UNION
SELECT 'x' AS $X
UNION
SELECT 'x' AS $X
```

The output:

```
{ X : 'x' },
{ Y : 1 }
```

Example 2

```
SELECT 1 as $Y
UNION
SELECT 'x' AS $X
UNION ALL
SELECT 'x' AS $X
```

The output:

```
{ X : 'x' },
{ Y : 1 },
{ X : 'x' }
```

Example 3

```
SELECT 1
UNION
SELECT 'x' AS $x
```

This returns an error because no catalog k-object or user-defined ad hoc is specified in the first SELECT query.

Example 4

```
SELECT {1 as $A, '2' AS $B} AS $C
UNION
SELECT {'2' AS $B, 1 as $A} AS $C
```

The output:

```
{ C : { A : 1, B : '2' } }
```

Example 5

```
SELECT {1 as $A, '2' AS $B} AS $C
UNION
SELECT {'2' AS $B, 1 as $A} AS $C2
```

The output:

```
{
    C2 : { A : 1, B : '2' }
},
{
    C : { A : 1, B : '2' }
}
```

INTERSECT

The INTERSECT of two stores produced by the respective SELECT statements is a store that includes all common host instances from the first store and the second store. All duplicates are eliminated. This operation is commutative – the result does not depend on the order of the operands.

Example

```
(SELECT 'x' AS $X UNION SELECT 'x' AS $Z)
INTERSECT
(SELECT 'x' AS $X UNION SELECT 'x' AS $Y)
```

The output is:

```
{ X : 'x' }
```

MINUS or EXCEPT

The MINUS, also called EXCEPT, of two stores produced by the respective SELECT statements is a store that includes all host instances from the first store that are not present in the second store. Al duplicates are eliminated. Like the minus operation in arithmetic, this operation is not commutative. The result depends on the order of the operands.

Example

```
(SELECT 'x' AS $X UNION SELECT 'x' AS $Z)
MINUS
(SELECT 'x' AS $X UNION SELECT 'x' AS $Y)
```

The output is:

```
{ z : 'x' }
```

3.12   User management statements

The user management statements are as follows.

CREATE USER
ALTER USER
SHOW USER
SHOW USERS
LIST USERS

CREATE PASSWORD_POLICY
ALTER PASSWORD_POLICY
DROP PASSWORD_POLICY
SHOW PASSWORD_POLICY
SHOW PASSWORD_POLICIES
LIST PASSWORD_POLICIES

CREATE USER

The CREATE USER statement creates a new KeySQL user login and account.

```
CREATE USER <login_name> WITH <comma_separated_option_list>

<option> ::=
    PASSWORD "=" ' <password> ' [ HASHED ] [ MUST_CHANGE ]
    | ENABLED "=" TRUE | FALSE
    | DEFAULT_SCHEMA "=" <schema_name>
    | CHECK_EXPIRATION "=" FALSE | TRUE
    | PASSWORD_POLICY "=" <policy_name>
    | USE_LDAP "=" FALSE | TRUE
    | LDAP_LINK "=" <ldap_link_name>
    | EXTENDED_PROPS "=" ' <json_document> '

<login_name> ::= <identifier>
<schema_name> ::= <identifier>
<policy_name> ::= <identifier>
<ldap_link_name> ::= <identifier>
```

`<login_name>` specifies the name of the login that is being created. There are two types of user logins: KeySQL server logins and LDAP logins.

Options:

- **PASSWORD** specifies the password for the user that is being created. Applies to KeySQL logins only for which the option is mandatory.
- **HASHED** specifies the hashed value of the password for the user that is being created.
- **MUST_CHANGE** specify that user must change the password on first log on. Password change procedure is specific for KeySQL client and may be implemented differently.
- **ENABLED** specifies that the created user is enables (by default) or disabled. Default value is **TRUE**.
- **DEFAULT_SCHEMA = `<schema_name>`** specifies default schema for the user that is being created. Default value is **PUBLIC**.
- **CHECK_EXPIRATION** specifies whether the user account should be checked for expiration. Expiration conditions are specified in the related password policy (see CREATE PASSWORD POLICY). Default value is **FALSE**.
- **PASSWORD_POLICY = `<policy_name>`** assign the password policy to the user account (see CREATE PASSWORD POLICY statement). If this option omitted, the **DEFAULT** policy is used.
- **USE_LDAP** specifies whether the user is linked to an LDAP account. The login name should correspond to the username in LDAP. Default value is **FALSE**.
- **LDAP_LINK = `<ldap_link_name>`** specify the LDAP server link to the user account

  LDAP links will be available in the future versions of KeySQL.

- **EXTENDED_PROPS** is a string representing a valid JSON document that may be useful for storing custom data.

Permissions

The statement requires CREATE_USER permission (server scope).

Examples

```
CREATE USER jsmith WITH
    PASSWORD = 'p@ssW0rd',
    DEFAULT_SCHEMA = sales_2020
;
```

```
CREATE USER "paul@keyark.com" WITH
    PASSWORD = 'abcd456',
    ENABLED = true,
    DEFAULT_SCHEMA = public,
    CHECK_EXPIRATION = true,
    PASSWORD_POLICY = default
;
CREATE USER mike WITH
    PASSWORD = '$67567A786F9D565BA786' HASHED MUST_CHANGE
;
CREATE USER "mike@orange.com" WITH
    USE_LDAP = true,
    LDAP_LINK = orange_ldap,
    EXTENDED_PROPS = '["Phone":"+33 123456789"]'
;
```

ALTER USER

The ALTER USER statement modifies an existing user account.

```
ALTER USER <login_name> WITH <comma_separated_option_list>

<option> ::=
    NAME "=" <login_name>
    | PASSWORD "=" ' <password> ' [ HASHED ] [ MUST_CHANGE ]
    | MUST_CHANGE_PASSWORD "=" FALSE | TRUE
    | ENABLED "=" TRUE | FALSE
    | DEFAULT_SCHEMA "=" <schema_name>
    | CHECK_EXPIRATION "=" FALSE | TRUE
    | PASSWORD_POLICY "=" <policy_name>
    | USE_LDAP "=" FALSE | TRUE
    | LDAP_LINK "=" <ldap_link_name>
    | EXTENDED_PROPS "=" ' <json_document> '
```

At least one specified option is required to run ALTER USER statement. The options are the same as for CREATE USER but also new ones:

- NAME specifies a new name for the user login. Use this option to rename a user.
- MUST_CHANGE_PASSWORD specify that user must change the password on the next log on.

> Changing an associated password policy will take an immediate effect on all related password checks.

Permissions

The statement requires ALTER_USER permission (server scope). A user can also modify their own properties without having explicit permission:

- PASSWORD
- EXTENDED_PROPS

Examples

```
ALTER USER jonh_s WITH NAME = jsmith;
ALTER USER jsmith WITH PASSWORD = 'Password' MUST_CHANGE;
ALTER USER jsmith WITH MUST_CHANGE_PASSWORD = true;
```

DROP USER

DROP USER statement deletes an existing user account.

```
DROP USER <login_name>
```

Deleting a user will also delete all user's permissions and exclude the user from all roles. If the user owned any objects like schemas or catalogs, the ownership will be "undefined". However, any user having the administrator privileges can take the ownership of these objects.

Permissions

The statement requires DROP_USER permission (server scope).

Examples

```
DROP USER jsmith;
DROP USER "jsmith@domain.com";
```

SHOW USER

The SHOW USER statement outputs the properties of an existing user account.

```
SHOW USER [ <login_name> ] [ WITH <comma_separated_option_list> ]

<option> ::=
    PWD_HASH "=" FALSE | TRUE
```

If login name is not specified, the statement outputs the current user account properties.

Options:

- PWD_HASH specifies whether the statement should print the password hash. The option may be useful for the user account export. The option requires ALL_USER permission.

  PWD_HASH option will be available in the future versions of KeySQL.

Permissions

The statement requires SHOW_USER permission (server scope). The current user can show their own properties without any explicit permissions.

To see the assigned password policy or LDAP link the user must have ALTER USER permission.

Output format

The statement supports both KeySQL and JSON output format. See also: SET PRETTY_PRINT, SET CURRENT_SCHEMA.

Examples.

The current user properties:

```
SET PRETTY_PRINT ON;
SHOW USER;
```

The result is as follows.

```
{
  user : {
    user_name : 'keysqluser',
    must_change_pwd : 0,
    enabled : 1,
    default_schema : 'KEYSQLUSER',
    check_expiration : 0,
    password_policy : {
      password_policy_name : 'DEFAULT',
      max_age : 90,
      max_age_unit : 1,
      min_length : 8,
      enforce_history : 0
    },
    use_ldap : 0,
    ldap_link : NULL,
```

```
    created_at : '2020-10-14 10:45:09.048095',
    updated_at : '2021-07-12 22:25:44.110148',
    pwd_updated_at : '2021-07-12 22:25:44.110147',
    extended_props :
'{"first_name":"Keysql","last_name":"Master","email":""}'
  }
}
```

Specific user properties:

```
SHOW USER jsmith;
```

Result:

```
{
  user: {
    user_name: 'jsmith',
    must_change_pwd: 0,
    enabled: 1,
    default_schema: 'TEST_DEV',
    check_expiration: 0,
    password_policy: {
      password_policy_name: 'CONTRACTORS',
      max_age: 90,
      max_age_unit: 1,
      min_length: 8,
      enforce_history: 0
    },
    use_ldap: 0,
    ldap_link: NULL,
    created_at: '2021-07-12 19:34:34.038645',
    updated_at: '2021-08-02 09:20:43.312105',
    pwd_updated_at: '2021-07-12 20:54:25.451295',
    extended_props:
'{"first_name":"John","last_name":"Smith","email":"js@corp.co"}'
  } }
```

SHOW USERS

The SHOW USERS statement outputs the properties of all existing user accounts.

```
SHOW USERS [ WITH <comma_separated_option_list> ]

<option> ::=
    PWD_HASH "=" FALSE | TRUE
```

Options:

- **PWD_HASH** specifies whether the statement should print password hash. The option may be useful for user account export. The option requires ALL_USER permission.

  > PWD_HASH option will be available in the future versions of KeySQL.

Permissions

The statement requires SHOW_USER permission (server scope).

Output format

The statement supports both KeySQL and JSON output format. See also: SET PRETTY_PRINT, SET CURRENT_SCHEMA.

Examples

```
SET PRETTY_PRINT ON;
SHOW USERS;
```

Result

```
{
  user : {
    user_name : 'keysqluser',
    must_change_pwd : 0,
    enabled : 1,
    default_schema : 'KEYSQLUSER',
    check_expiration : 0,
    password_policy : {...},
    use_ldap : 0,
    ldap_link : NULL,
    created_at : '2020-10-14 10:45:09.048095',
    updated_at : '2021-07-12 22:25:44.110148',
    pwd_updated_at : '2021-07-12 22:25:44.110147',
    extended_props :
'{"first_name":"Keysql","last_name":"Master","email":""}'
  }
},
{
  user: {
    user_name: 'jsmith',
    must_change_pwd: 0,
    enabled: 1,
    default_schema: 'PUBLIC',
    check_expiration: 0,
    password_policy: {...},
    use_ldap: 0,
    ldap_link: NULL,
```

```
        created_at: '2021-05-12 20:15:31.748282',
        updated_at: '2021-05-18 13:55:49.181589',
        pwd_updated_at: '2021-05-12 20:16:25.917191',
        extended_props: '{"email":"test@keyark.com", "age":31}'
    }
}
```

LIST USERS

The LIST USERS statement returns the lexicographically ordered list of all user
login names.

```
LIST USERS
```

Permissions

The statement requires LIST_USERS permission.

Example

```
LIST USERS
```

The output:

```
{USER_NAME:'keysqladmin'},
{USER_NAME:'keysqluser'}
```

CREATE PASSWORD_POLICY

The CREATE PASSWORD_POLICY statement creates a new policy to manage
passwords of user accounts.

```
CREATE PASSWORD_POLICY <policy_name> WITH
<comma_separated_option_list>

<options> ::=
  MAX_AGE "=" <age> [ DAYS ]
  | MIN_LENGTH "=" <min_length>
  | ENFORCE_HISTORY "=" FALSE | TRUE
<age> ::= 1..999
<min_length> ::= 1..14
```

The <policy_name> specifies the name of the policy being created. The name
then will be used in CREATE USER and ALTER USER statements.

Options:

- `MAX_AGE` determines the period (since the last password change) when a password can be used before the system requires the user to change it. The option is required, the default period unit is DAYS.
- `MIN_LENGTH` specifies the minimal number of characters that can make up a password for a user account. The option is required, and its value should be between 4 and 14 characters.
- `ENFORCE_HISTORY` specifies the number of the consecutive passwords for a user account before an old password can be reused.

> ENFORCE_HISTORY option will be available in the future versions of KeySQL.

Permissions

The statement requires `CREATE_PASSWORD_POLICY` permission (server scope).

Examples

```
CREATE PASSWORD_POLICY contractors WITH
    MAX_AGE = 90 DAYS,
    MIN_LENGTH = 8
```

DEFAULT password policy

The password policy named "DEFAULT" is the system object that is used by any KeySQL user by default when not specified.

> Note that the name DEFAULT cannot be altered or deleted.

ALTER PASSWORD_POLICY

The ALTER PASSWORD_POLICY statement modifies an existing password policy.

```
ALTER PASSWORD_POLICY <policy_name> WITH
<comma_separated_option_list>

<options> ::=
  NAME "=" <policy_name>
  | MAX_AGE "=" <age> [ DAYS ]
  | MIN_LENGTH "=" <min_length>
  | ENFORCE_HISTORY "=" FALSE | TRUE
<age> ::= 1..999
<min_length> ::= 1..14
```

The <policy_name> specifies the name of the existing policy. The options are the same as for CREATE PASSWORD_POLICY with one addition:

- `NAME` specifies a new name. Use this option to rename a password policy.

  Changing an associated password policy will take an immediate effect on password checks for all related users.

Permissions

The statement requires `ALTER_PASSWORD_POLICY` permission (server scope).

Examples

```
ALTER PASSWORD_POLICY contractors WITH
    NAME = short_contractors
    MAX_AGE = 30 DAYS,
    MIN_LENGTH = 8
```

DROP PASSWORD_POLICY

DROP PASSWORD_POLICY statement deletes an existing password policy.

```
DROP PASSWORD_POLICY <policy_name>
```

  You cannot drop a policy related to one or more existing users; the corresponding error will be returned.

To see related users, use SHOW PASSWORD_POLICY statement.

Permissions

The statement requires `DROP_PASSWORD_POLICY` permission (server scope).

Examples

```
DROP PASSWORD_POLICY contractors
```

SHOW PASSWORD_POLICY

The SHOW PASSWORD_POLICY statement outputs the properties of an existing password policy.

```
SHOW PASSWORD_POLICY [ <policy_name> ] [ WITH
<comma_separated_option_list> ]

<option> ::=
    USAGE
```

Options:

- **USAGE** specifies if the statement should print related user login names

Permissions

The statement requires SHOW_PASSWORD_POLICY permission (server scope).

Output format

The statement supports both KeySQL and JSON output format.

The max_age_unit item value correspondences:

- 1 – DAYS (by default).
- other values will be available in the future versions of KeySQL.

Examples

```
SHOW PASSWORD_POLICY contractors WITH USAGE;
```

Result:

```
{
    password_policy: {
        password_policy_name: 'CONTRACTORS',
        max_age: 90,
        max_age_unit: 1,
        min_length: 8,
        enforce_history: 0,
        usages: {
            user_name: 'JSmith',
            user_name: 'Wendy'
        }
    }
}
```

SHOW PASSWORD_POLICIES

The SHOW PASSWORD_POLICIES statement outputs the properties of all existing password policies.

```
SHOW PASSWORD_POLICIES [ WITH <comma_separated_option_list> ]

<option> ::=
    USAGE
```

Options:

- **USAGE** specifies whether the statement should print related user login names.

Permissions

The statement requires SHOW_PASSWORD_POLICY permission (server scope).

Output format

The statement supports both KeySQL and JSON output format.

Examples

```
SET PRETTY_PRINT ON;
SHOW PASSWORD_POLICIES;
```

Result

```
{
    password_policy: {
        password_policy_name: 'CONTRACTORS',
        max_age: 90,
        max_age_unit: 1,
        min_length: 8,
        enforce_history: 0,
        usages: {
            user_name: 'JSmith',
            user_name: 'Wendy'
        }
    }
},
{
    password_policy: {
        password_policy_name: 'EMPLOYEES',
        max_age: 120,
        max_age_unit: 1,
        min_length: 7,
        enforce_history: 0,
        usages: {
            user_name: 'Emp01',
            user_name: 'Emp02',
            user_name: 'Emp03'
        }
    }
}
```

LIST PASSWORD_POLICIES

The LIST USERS statement returns the lexicographically ordered list of all user login names.

```
LIST PASSWORD_POLICIES
```

Permissions

The statement requires LIST_PASSWORD_POLICIES permission.

Example

```
LIST PASSWORD_POLICIES
```

The output:

```
{PASSWORD_POLICY_NAME:'contractors'},
{PASSWORD_POLICY_NAME:'default'}
```

3.13   Access control statements

The access control statements are as follows.

GRANT
REVOKE
SHOW PERMISSION DEFINITIONS
SHOW PERMISSIONS

GRANT

The GRANT statement assigns the specified permission on the securable to the principal.

```
GRANT <permission_list> [ ON <securable> ] TO <principal>

<permission_list> ::= <permission_name> [ "," <permission_list> ]
<permission_name> ::=
   ALL
   | <securable_specific_permission>

<securable> ::=
   SERVER
   | SCHEMA <schema_name>
   | CATALOG <qualified_catalog_name>
```

```
| STORE <qualified_store_name>
| USER <user_name>
| PASSWORD_POLICY <password_policy_name>

<principal> ::= <user_name> | <role_name>
```

Roles will be available in the future versions of KeySQL.

The SERVER securable is not required in ON clause.

The securable specific permissions are as follows.

- STORE (securable and scope)
  - ALL
  - ALTER STORE
  - DELETE
  - DROP STORE
  - INSERT
  - LIST STORE
  - SELECT
  - SHOW STORE
  - UPDATE
- CATALOG
  - ALL
  - ALL KEYOBJECT
  - ALL STORE
  - ALTER CATALOG
  - ALTER KEYOBJECT
  - CREATE KEYOBJECT
  - CREATE STORE
  - DROP CATALOG
  - DROP KEYOBJECT
  - LIST CATALOG
  - LIST KEYOBJECT
  - SHOW CATALOG
  - SHOW KEYOBJECT
  - All permissions of STORE scope
- SCHEMA
  - ALL
  - ALL CATALOG
  - ALL STORE

- ALTER SCHEMA
- CREATE CATALOG
- LIST SCHEMA
- SHOW SCHEMA
- All permissions of CATALOG scope
- All permissions of STORE scope
- SERVER
  - ALL
  - ALL PASSWORD_POLICY
  - ALL SCHEMA
  - ALL USER
  - ALTER PASSWORD_POLICY
  - ALTER SCHEMA
  - ALTER USER
  - CREATE PASSWORD_POLICY
  - CREATE SCHEMA
  - CREATE USER
  - DROP PASSWORD_POLICY
  - DROP SCHEMA
  - DROP USER
  - LIST PASSWORD_POLICY
  - LIST SCHEMA
  - LIST USER
  - SHOW PASSWORD_POLICY
  - SHOW PERMISSIONS
  - SHOW SCHEMA
  - SHOW USER

Permissions

The statement requires the ownership on the specified securable or the membership in server administrator role.

Examples

Allow the user to manage other users:

```
GRANT ALL USER TO jsmith;
GRANT ALL PASSWORD_POLICY to jsmith;
```

Allow the user to modify any store (including the ones that will be created) in the schema:

```
GRANT INSERT, UPDATE, DELETE ON SCHEMA MY_SCHEMA TO wendy
```

Allow the user to read from the store:

```
GRANT SELECT ON STORE hr.employees TO jsmith
```

Allow the user to manage the schema:

```
GRANT ALL ON SCHEMA sales TO mike
```

Allow the user to manage the catalog k-objects:

```
GRANT ALL KEYOBJECT ON CATALOG hr.management TO katty
```

REVOKE

The REVOKE statement removes the specified permission on the securable from the principal.

```
REVOKE <permission_list> [ ON <securable> ] FROM <principal> [
CASCADE | ATOMIC ]

<permission_list> ::= <permission_name> [ "," <permission_list> ]
<permission_name> ::=
    ALL
    | <securable_specific_permission>

<securable> ::=
    SERVER
    | SCHEMA <schema_name>
    | CATALOG <qualified_catalog_name>
    | STORE <qualified_store_name>
    | USER <user_name>
    | PASSWORD_POLICY <password_policy_name>

<principal> ::= <user_name> | <role_name>
```

> Roles will be available in the future versions of KeySQL.

The SERVER securable is not required in ON clause.

The securable specific permissions are the same as described in the GRANT statement.

Options:

- **CASCADE** mode (by default) revokes the permission for the specified securable and for all its child ones. For example, revoking UPDATE on a catalog leads to revoking UPDATE on all stores defined for this catalog.

- ATOMIC mode revokes the permission only on the specified securable.

Permissions

The statement requires the ownership on the specified securable or the membership in server administrator role.

Examples

Deny modifying a store for a user:

```
REVOKE INSERT, DELETE, UPDATE ON STORE hr.employees FROM jsmith
```

Allow the user to modify only the catalog:

```
-- previously granted
-- GRANT ALL ON CATALOG hr.management TO katty
REVOKE ALL STORE ON CATALOG hr.management FROM katty
```

Deny managing stores excluding existing ones

```
-- previously granted
-- GRANT ALL STORE ON SCHEMA hr TO jsmith
REVOKE ALL STORE ON SCHEMA hr FROM jsmith ATOMIC
```

SHOW PERMISSION DEFINITIONS

The SHOW PERMISSION DEFINITIONS statement shows all existing permissions. The output format is export friendly.

```
SHOW PERMISSION DEFINITIONS
```

Permissions

None

Example

Output:

```
{
    PERMISSION_DEFINITION: {
        SECURABLE: 'SERVER',
        PERMISSION: 'ALL',
        COMPONENT_PERMISSIONS: {
            PERMISSION: 'ALL PASSWORD_POLICY',
```

```
                   PERMISSION: 'ALL SCHEMA',
                   PERMISSION: 'ALL USER'
            }
        }
    },
    {
        PERMISSION_DEFINITION: {
            SECURABLE: 'SERVER',
            PERMISSION: 'ALL USER',
            COMPONENT_PERMISSIONS: {
                   PERMISSION: 'ALTER USER',
                   PERMISSION: 'CREATE USER',
                   PERMISSION: 'DROP USER',
                   PERMISSION: 'LIST USER',
                   PERMISSION: 'SHOW PERMISSIONS',
                   PERMISSION: 'SHOW USER'
            }
        }
    },
    {
        PERMISSION_DEFINITION: {
            SECURABLE: 'SERVER',
            PERMISSION: 'LIST USER',
            COMPONENT_PERMISSIONS: NULL
        }
    },
    ...
```

SHOW PERMISSIONS

The SHOW PERMISSIONS statement displays user grants and revokes.

```
SHOW PERMISSIONS [ FOR ( USER | ROLE ) <name> ]
```

Statement without specified user or role name will show current user permissions.

Permissions

The current user does not require any permissions to see their own authorizations, otherwise the statement requires SHOW_PERMISSIONS.

NONE permission

The 'NONE' permission may appear in the statement output. Usually this means that previously granted permission has been revoked.

Example

```
SHOW PERMISSIONS FOR USER bgates
```

Result:

```
{
  USER_PERMISSIONS : {
    GRANTS : {
      GRANT : {
        NAME : 'BGATES',
        PERMISSIONS : {
          PERMISSION : 'ALL'
        },
        SECURABLE : 'SCHEMA'
      },
      GRANT : {
        NAME : 'PUBLIC',
        PERMISSIONS : {
          PERMISSION : 'ALL'
        },
        SECURABLE : 'SCHEMA'
      },
      GRANT : {
        NAME : 'ABS',
        PERMISSIONS : {
          PERMISSION : 'NONE'
        },
        SECURABLE : 'SCHEMA'
      }
    },
    NAME : 'BGATES',
    TYPE : 'user'
  }
}
```

Server and session management   4

4.1   Session level statements

Session level statements show or modify settings for the current user session only. Disconnect and the following connect will reset all parameters to their default values.

> Some KeySQL clients may override the default values of session state when connecting.

The session level statements are as follows.

SET CURRENT_SCHEMA
SET OUTPUT_FORMAT
SET PRETTY_PRINT

SET CURRENT_SCHEMA

The SET CURRENT_SCHEMA statement shows or sets the KeySQL schema used by default in the current session.

```
SET CURRENT_SCHEMA [ <schema_name> ]
```

SET CURRENT_SCHEMA without parameters returns the current value.

Permissions

None

Examples

```
SET CURRENT_SCHEMA;
```

Result:

```
{current_schema:'SYSTEM'}
```

Set the value and list the catalogs.

```
SET CURRENT_SCHEMA public;
LIST CATALOGS; -- all catalogs from public schema
```

SET OUTPUT_FORMAT

The SET OUTPUT_FORMAT statement defines which format should be used when server sends a reply to a client.

```
SET OUTPUT_FORMAT [ ( KEYSQL | JSON ) ]
```

Default value is KEYSQL (KeySQL Object Notation). SET OUTPUT_FORMAT without parameters returns the current value.

Permissions

None

Examples

```
SET OUTPUT_FORMAT JSON;
SET OUTPUT_FORMAT;
```

Result:

```
{"output_format":"JSON"}
```

SET PRETTY_PRINT

The SET PRETTY_PRINT statement specifies whether the server output will be formatted as human readable including indents and line feeds.

```
SET PRETTY_PRINT [ ( ON | OFF ) ]
```

Default value is OFF. SET PRETTY_PRINT without parameters returns the current value.

Permissions

None

Examples

Show current value:

```
SET PRETTY_PRINT
```

Result:

```
{"pretty_print":"off"}
```

Change setting:

```
SET PRETTY_PRINT ON;
SET PRETTY_PRINT;
```

Result:

```
{
    "pretty_print": "on"
}
```

4.2   Server level statements

The server level statements are as follows.

EXPLAIN

SHOW SERVER_INFO

SHOW VERSION

UPDATE ATATISTICS

EXPLAIN

The EXPLAIN statement shows the generated KeySQL statement structure.

```
EXPLAIN <explain_mode> [ <KeySQL_statement> ]

<explain_mode> ::= GENERATED
<KeySQL_statement> ::= any valid KeySQL statement
```

> The statement is under construction and may not work with some statements.

Permissions

The statement requires same permission as for the KeySQL statement.

Examples

```
SET PRETTY_PRINT ON;
EXPLAIN GENERATED SHOW USERS;
```

Returns

```
{
  explain : {
    {
      Step# : '1',
      Level : '0',
      SQL : ''
    }
  }
}
```

SHOW SERVER_INFO

The SHOW SERVER_INFO statement returns the set of values related to the KeySQL server state.

```
SHOW SERVER_INFO
```

Permissions

The statement requires KeySQL administrator permissions.

Output format

The statement returns several k-objects describing KeySQL server state

K-object	Value	Description
SERVER_INFO	SERVER_STARTED	Date and time of the last KeySQL server start
	SERVER_ACTIVE	Time since the server has been started
	CONNECTION	Number of active connections
LIMITS	MAX_MEMORY	Maximum amount of RAM that can be allocated by server
	MAX_QUERY_MEMORY	Maximum amount of RAM that can be allocated per query
	MAX_CONNECTIONS	Maximum number of concurrent connections

	MAX_STATEMENTS	Maximum number of queries which are running or queued. The server returns "server is busy" error after this count is reached
	MAX_ONGOING_STATEMENTS	Maximum number of queries in progress
	WAITED_TIMEOUT_SEC	Wait timeout for a query. "0" means "waiting indefinitely"
REQUESTS	PROCESSED	Total number of processed queries
	ONGOING	Number of currently running queries
	DROPPED	Number of cancelled queries with "server busy" error
	WAITED	Number of queued queries. The number is always between MAX_ONGOING_STATEMENTS and MAX_STATEMENTS
	AVG_ELAPSED_TIME	Average elapsed time per query
	MAX_ELAPSED_TIME	Maximum elapsed time per query
	WORKERS_QUEUE_SIZE	Number of queries in waiting for an available thread (CPU/core)

	SQL_QUEUE_SIZE	Number of queries in waiting for an available SQL connection
CPU_SEC	USER_MODE	Total CPU usage in user mode, seconds
	KERNEL_MODE	Total CPU usage in kernel mode, seconds
MEMORY	VIRTUAL_MEMORY	Shows the amount of virtual memory in usage
	VIRTUAL_MEMORY_PEAK	Shows the peak of usage of virtual memory
	PHYSICAL_MEMORY	Shows the amount of physical memory in usage
	PHYSICAL_MEMORY_PEAK	Shows the peak of usage of physical memory
	SWAP	Shows the amount of memory used in swap
PROCESS_STATUS	THREADS	Total number of threads owned by server process
	VOLUNTARY_CONTEXT_SWITCHES	Number of times that the program was context-switched voluntarily, for instance while waiting for an I/O operation to complete
	NONVOLUNTARY_CONTEXT_SWITCHES	Number of times the process was context-switched involuntarily (because the time slice expired)

Examples

```
SHOW SERVER_INFO
```

Returns:

```
{
  SERVER_INFO : {
    SERVER_STARTED : '2022-03-18 18:17:18.546027',
    SERVER_ACTIVE : '9 days 03:37:51.602343',
    CONNECTION : 43,
    LIMITS : {
      MAX_MEMORY : 'unlimited',
      MAX_QUERY_MEMORY : '10240000 kB',
      MAX_CONNECTIONS : 65535,
      MAX_STATEMENTS : 2147483647,
      MAX_ONGOING_STATEMENTS : 2147483647,
      WAITED_TIMEOUT_SEC : 0
    },
    REQUESTS : {
      PROCESSED : 15193,
      ONGING : 1,
      DROPPED : 0,
      WAITED : 0,
      AVG_ELAPSED_TIME : 0.0245021213948527,
      MAX_ELAPSED_TIME : 1.558439305,
      WORKERS_QUEUE_SIZE : 0,
      SQL_QUEUE_SIZE : 0
    },
    CPU_SEC : {
      USER_MODE : 233.248662,
      KERNEL_MODE : 28.676554
    },
    MEMORY : {
      VIRTUAL_MEMORY : '5152964 kB',
      VIRTUAL_MEMORY_PEAK : '5157576 kB',
      PHYSICAL_MEMORY : '507852 kB',
      PHYSICAL_MEMORY_PEAK : '507852 kB',
      SWAP : '0 kB'
    },
    PROCESS_STATUS : {
      THREADS : 68,
      VOLUNTARY_CONTEXT_SWITCHES : 990,
      NONVOLUNTARY_CONTEXT_SWITCHES : 4
    }
  }
}
```

SHOW VERSION

The SHOW VERSION statement returns the `version_info` composition k-object instance containing information about the running KeySQL server instance.

```
SHOW VERSION
```

The `version_info` k-object has following elements:

- `product_name` : the product name.
- `version_num` : product major and minor version, and the revision number separated by dots.
- `release_date` : the date and the time of product release.
- `build_num` : internal build information which is useful for user feedbacks.
- `default_dbms`: the type and dialect of the storage layer database system.

The `version_info` k-object can also contain other k-objects that will be added in the future versions of KeySQL.

Permissions

None

Example

```
SHOW VERSION
```

Result:

```
{
  version_info : {
    product_name : 'KeySQL Server',
    version_num : '2.20.0',
    release_date : '2023-12-20 22:07:51',
    build_num : '168 (master:1f05a6f)'
    default_dbms: 'Provider: PostgreSQL, SQL dialect: PostgreSQL
  }
}
```

UPDATE STATISTICS

The UPDATE STATISTICS statement is useful for underlying storage SQL DBMS having simplified or manual statistics update mode. When a large pack of data has been inserted into a store, the statistics can be updated manually to improve the storage SQL DBMS query execution.

```
UPDATE STATISTICS [ <qualified_store_name> ]
```

UPDATE STATISTICS will update statistics for all tables in the database. When a store name is specified, the statement will update statistics for the store underlying tables only.

Permissions

UPDATE STATISTICS requires server scope privilege CREATE SCHEMA.

UPDATE STATISTICS with the specified store name requires privileges to modify the store data: INSERT, UPDATE, and DELETE.

Examples

```
UPDATE STATISTICS;
UPDATE STATISTICS sales_2020.shop_012;
```

KeySQL functions and operators

5

The examples below are based on `BLRD_STORE` store introduced in the first chapter with the following additional k-objects.

```
CREATE KEYOBJECT num_val NUMBER IN CATALOG blrd_catalog;
CREATE KEYOBJECT date_val DATE IN CATALOG blrd_catalog;
CREATE KEYOBJECT str_val CHAR IN CATALOG blrd_catalog;
CREATE KEYOBJECT num_pair {num, num_val} IN CATALOG blrd_catalog;
```

5.1   Aggregate functions

COUNT

The COUNT function returns the number of items.

```
COUNT( * | <qualified_keyobject_name> )
```

Example

```
SELECT COUNT(*) FROM blrd_store
;
SELECT COUNT(DISTINCT ball.num) FROM blrd_store;
```

Those statements produce the following respective results.

```
{
  COUNT : 2
}
{
  COUNT_BALL_NUM : 15
}
```

AVG

The AVG function returns the average of numeric k-object values. All NULL values are considered as 0 when calculating.

```
AVG(<qualified_keyobject_name>)
```

Example

```
SELECT AVG(ball.num) FROM blrd_store
```

Returns

```
{
    AVG_BALL_NUM : 8
}
```

SUM

The SUM function returns the sum of numeric k-object values. All NULL values are considered as 0 when calculating.

```
SUM(<qualified_keyobject_name>)
```

Example

```
SELECT SUM(ball.num) FROM blrd_store
```

Returns

```
{
    SUM_BALL_NUM : 120
}
```

MIN

The MIN function returns the smallest value of numeric k-objects.

```
MIN(<qualified_keyobject_name>)
```

All NULL values are ignored when comparing values. When only NULL values are selected, the result is NULL, too.

Example

```
SELECT MIN(ball.num) FROM blrd_store;
SELECT MIN(ball.num) FROM blrd_store WHERE ball.num IS NULL;
```

Those statements produce the following respective results.

```
{
    MIN_BALL_NUM : 1
}
```

```
{
    MIN_BALL_NUM : NULL
}
```

MAX

The MAX function returns the largest value of numeric k-objects.

```
MAX(<qualified_keyobject_name>)
```

All NULL values are ignored when comparing values. When only NULL values are selected, the result is NULL, too.

Example

```
SELECT MAX(ball.num) FROM blrd_store;
SELECT MAX(ball.num) FROM blrd_store WHERE ball.num IS NULL;
```

Those statements produce the following respective results.

```
{
    MAX_BALL_NUM : 15
}
```

```
{
    MAX_BALL_NUM : NULL
}
```

MEDIAN

The median is the number in the middle of a set of numbers.

The MEDIAN function calculates median value from a set of numeric k-object values.

```
MEDIAN(<expression>)
```

Note that keyobjects in the expression should be siblings.

Example

```
SELECT MEDIAN(ball.num)
FROM blrd_store
```

Returns

```
{
    MEDIAN_BALL_NUM : 8
}
```

MULT

The MULT function returns the product of numeric k-object values. All NULL values are ignored.

```
MULT(<qualified_keyobject_name>)
```

All NULL values are skipped when multiplying values. When only NULL values are selected, the result is NULL.

Example

```
SELECT MULT(ball.num) FROM blrd_store
;
SELECT MULT(ball.num) FROM blrd_store
WHERE ball.num IS NULL OR ball.num < 3
;
SELECT MULT(ball.num) FROM blrd_store
WHERE ball.num IS NULL;
```

Those statements produce the following respective results.

```
{
    MULT_BALL_NUM : 1307674368000
}
{
    MULT_BALL_NUM : 2
}
{
    MULT_BALL_NUM: NULL
}
```

PERCENTILE

The PERCENTILE function computes a specified percentile of the ascending sorted values of any given numeric k-object for all entries in the group.

PERCENTILE() is an inverse distribution function that assumes a continuous distribution model. The MEDIAN function is a specific case of PERCENTILE where the percentile value defaults to 0.5.

```
PERCENTILE(<expression>, <const expression>)
```

Arguments:

- `<expression>` is any numeric k-object expression with the same parent. Note that keyobjects in the expression should be siblings.
- `<const expression>` is a NUMBER expression from interval [0.0,1.0]

The function returns NUMBER.

Example

```
SELECT PERCENTILE(ball.num, 0.5) AS $perc
FROM blrd_store
```

Returns

```
{
  PERC : 8
}
```

STDDEV_POP

Standard deviation is a measure of how spread out the values in a data set are. The standard deviation shows how much variation exists from the average (mean).

If all values in the data set are taken into the calculation, this standard deviation is called population standard deviation.

The STDDEV_POP function calculates population standard deviation from a set of numeric k-object values.

```
STDDEV_POP(<expression>)
```

Note that keyobjects in the expression should be siblings.

Example

```
SELECT STDDEV_POP(ballset.diam) AS $stdpop
FROM blrd_store;
```

Returns

```
{
  STDPOP : 2.175
}
```

STDDEV_SAMP

If a subset of values or a sample is taken into the calculation, this standard deviation is called sample standard deviation.

The STDDEV_SAMP function calculates sample standard deviation from a set of numeric k-object values.

```
STDDEV_SAMP(<expression>)
```

Note that keyobjects in the expression should be siblings.

Example

```
SELECT STDDEV_SAMP(ballset.diam) AS $stdsamp
FROM blrd_store
```

Returns

```
{
   STDSAMP : 3.07591449816148
}
```

VAR_POP

The variance is a measure of how far a set of numbers is spread out. It is one of several descriptors of a probability distribution.

The VAR_POP function calculates compute statistical variance for an entire population data (a set of numeric k-object values).

```
VAR_POP(<expression>)
```

Note that keyobjects in the expression should be siblings.

Example

```
SELECT VAR_POP(ballset.diam) AS $varpop
FROM blrd_store
```

Returns

```
{
   VARPOP : 4.730625
}
```

VAR_SAMP

The VAR_POP function calculates statistical variance from sample data (a set of numeric k-object values).

```
VAR_SAMP(<expression>)
```

Note that keyobjects in the expression should be siblings.

Example

```
SELECT VAR_SAMP(ballset.diam) AS $varsamp
FROM blrd_store
```

Returns

```
{
  VARSAMP : 9.46125000000001
}
```

5.2   Arithmetic operators

Arithmetic operators have two numeric or integer expressions as the operands.

Return types:

- Numeric when the first or both operands are numeric.
- Integer when the first or both operands are integer.
- NULL when one of the operands is null.

+ (Addition)

Adds two numbers.

```
expression + expression
```

Example 1

```
SELECT 5 + 3 AS $I;
SELECT 5.5 + 2 AS $N;
SELECT NULL + 2 AS $X;
```

Returns

```
{I:8}
```

```
{N:7.5}
{X:NULL}
```

- (Subtraction)

Subtracts one numeric expression from another.

```
expression - expression
```

Example

```
SELECT 5 - 3 AS $I;
SELECT 5.5 - 2 AS $N;
SELECT 2 - NULL AS $X;
```

Returns

```
{I:2}
{N:3.5}
{X:NULL}
```

* (Multiplication)

Multiplies two numeric expressions.

```
expression * expression
```

Example

```
SELECT 5 * 3 AS $I;
SELECT 5.5 * 2 AS $N;
SELECT 2 * NULL AS $X;
```

Returns

```
{I:15}
{N:11}
{X:NULL}
```

/ (Division)

Divides one numeric expression by another.

```
dividend / divisor
```

Example 1

```
SELECT 5 / 3 AS $I;
SELECT 5.5 / 2 AS $N;
SELECT NULL / 2 AS $X;
```

Returns

```
{I:1}
{N:2.75}
{X:NULL}
```

Example 2

```
SELECT ballset.diam / 100 AS num_val
FROM blrd_store
```

Returns

```
{NUM_VAL:0.615},
{NUM_VAL:0.5715},
{NUM_VAL:0.68},
{NUM_VAL:0.525}
```

% (Modulus)

Returns the remainder of one number divided by another number.

```
dividend % divisor
```

Example

```
SELECT 5 % 3 AS $I;
SELECT 5.5 % 2 AS $N;
SELECT NULL % 2 AS $X;
```

Returns

```
{I:2}
{N:1.5}
{X:NULL}
```

5.3 Bitwise operators

Bitwise operators perform operations on one or two expressions (operands) of integer data type.

Return types by default:

- Integer when both operands are not null.
- NULL when one operand is null.

& (Bitwise AND)

Performs a bitwise logical AND operation between two integer values.

```
expression & expression
```

Example

```
SELECT 6 & 2 AS $I;
SELECT 4 & NULL AS $I;
```

Returns

```
{I:2}
{I:NULL}
```

| (Bitwise OR)

Performs a bitwise logical OR operation between two integer values.

```
expression | expression
```

Example

```
SELECT 4 | 2 AS $I;
SELECT 4 | NULL AS $I;
```

Returns

```
{I:6}
{I:NULL}
```

^ (Bitwise exclusive OR)

Performs a bitwise exclusive OR operation between two integer values.

```
expression ^ expression
```

Example

```
SELECT 6 ^ 10 AS $I
SELECT 4 ^ NULL AS $I
```

Returns

```
{I:10}
{I:NULL}
```

~ (Bitwise NOT)

Performs a bitwise logical NOT operation on an integer value.

```
~ expression
SELECT ~2 AS $I;
SELECT ~(-2) AS $I;
SELECT ~NULL AS $I;
SELECT ~(1 + NULL) AS $I;
```

Example

Returns

```
{I:-3}
{I:1}
{I:NULL}
{I:NULL}
```

>> (Bitwise shift right)

Shifts the bits in the first operand to the right by N bits (second operand).

```
expression >> expression
```

Example

```
SELECT 1 >> 1 AS $I;
SELECT 4 >> 2 AS $I;
SELECT NULL >> 1 AS $I;
```

Result

```
{I:0}
{I:1}
```

```
{I:NULL}
```

<< (Bitwise shift left)

Shifts the bits in the first operand to the left by N bits (second operand).

```
expression << expression
```

Example

```
SELECT 1 << 1 AS $I;
SELECT 4 << 2 AS $I;
SELECT NULL << 1 AS $I;
```

Result

```
{I:2}
{I:16}
{I:NULL}
```

5.4   Mathematical functions

ABS

The ABS function returns the absolute (positive) value of the specified numeric expression. Hence, the function changes negative values to positive values, and has no effect on zero or positive values.

```
ABS(<numeric_expression>)
```

Arguments

Any expression returning numeric value.

Return types

Numeric

Example

```
SELECT ABS(60 - ballset.diam) AS num_val
FROM blrd_store
```

Returns

```
{NUM_VAL: 1.5},
{NUM_VAL: 2.85},
```

ACOS

The ACOS function returns the angle, in radians, whose cosine is the specified.

```
ACOS(<numeric_expression>)
```

Arguments

Any expression returning numeric value.

Return types

Numeric

Example

```
SELECT ACOS(ball.num - 1) * 2 AS num_val
FROM blrd_store
WHERE ball.num = 1
```

Returns

```
{
   NUM_VAL : 3.14159265358979
}
```

ASIN

The ASIN function returns the angle, in radians, whose sine is specified.

```
ASIN(<numeric_expression>)
```

Arguments

Any expression returning numeric value.

Return types

Numeric

Example

```
SELECT ASIN(ball.num) * 2 AS num_val
FROM blrd_store
WHERE ball.num = 1
```

Returns

```
{
  NUM_VAL : 3.14159265358979
}
```

ATAN

The ATAN function returns the angle, in radians, whose tangent is specified.

```
ATAN(<numeric_expression>)
```

Arguments

Any expression returning numeric value.

Return types

Numeric

Example

```
SELECT ATAN(PI() / ball.num) AS num_val
FROM blrd_store
WHERE ball.num = 2
```

Returns

```
{
  NUM_VAL : 1.00388482185389
}
```

ATAN2

The ATAN function returns the angle, in radians, between the positive X-axis and the ray from the origin to the specified point (y, x).

```
ATAN2(<y_value>, <x_value>)
```

Arguments

X and Y are any expression returning numeric values.

Return types

Numeric

Example

```
SELECT ATAN2(ball.num, ball.num) AS num_val
FROM blrd_store
WHERE ball.num = 1
```

Returns

```
{
  NUM_VAL : 0.785398163397448
}
```

CBRT

The CBRT function returns the cubic root of the specified value.

```
CBRT(<numeric_expression>)
```

Arguments

Any expression returning numeric value.

Return types

Numeric

Example

```
SELECT { ball.num, CBRT(ball.num) AS num_val } AS num_pair
FROM blrd_store
WHERE ball.num < 4
```

Returns

```
{
  NUM_PAIR : {
    NUM : 1,
    NUM_VAL : 1
  }
},
{
  NUM_PAIR : {
    NUM : 2,
    NUM_VAL : 1.25992104989487
  }
},
{
  NUM_PAIR : {
```

```
    NUM : 3,
    NUM_VAL : 1.44224957030741
  }
}
```

CEIL

The CEIL function returns the smallest integer greater than, or equal to, the specified numeric expression.

```
CEIL(<numeric_expression>)
```

Arguments

Any expression returning numeric value.

Return types

Numeric (integer)

Example

```
SELECT { ball.num, CEIL(ball.num / 2) AS num_val } AS num_pair
FROM blrd_store
WHERE ball.num < 5
```

Returns

```
{
  NUM_PAIR : {
    NUM : 1,
    NUM_VAL : 1
  }
},
{
  NUM_PAIR : {
    NUM : 2,
    NUM_VAL : 1
  }
},
{
  NUM_PAIR : {
    NUM : 3,
    NUM_VAL : 2
  }
},
{
  NUM_PAIR : {
```

```
    NUM : 4,
    NUM_VAL : 2
  }
}
```

COS

The COS function returns the trigonometric cosine of the specified angle in radians.

```
COS(<numeric_expression>)
```

Arguments

Any expression returning numeric value.

Return types

Numeric

Example

```
SELECT COS(PI() / ball.num) AS num_val, ball.num
FROM blrd_store
WHERE ball.num < 3
```

Returns

```
{
  RESULT : {
    #NUM_VAL : {
      NUM_VAL : -1,
      NUM_VAL : 6.12323399573677e-17
    },
    #NUM : {
      NUM : 1,
      NUM : 2
    }
  }
}
```

COT

The COT function returns the trigonometric cotangent of the specified angle in radians.

```
COT(<numeric_expression>)
```

Arguments

Any expression returning numeric value.

Return types

Numeric

Example

```
SELECT COT(PI() / ball.num) AS num_val
FROM blrd_store
WHERE ball.num = 2
```

Returns

```
{
    NUM_VAL : 6.12323399573677e-17
}
```

EXP

The EXP function returns exponential (e^x) value of the specified expression.

```
EXP(<numeric_expression>)
```

The constant e (2.718281...), is the base of natural logarithms.

Arguments

Any expression returning numeric value.

Return types

Numeric

Example

```
SELECT EXP(ballset.diam / 10) AS num_val
FROM blrd_store
LIMIT 1
```

Returns

```
{NUM_VAL: 468.717386782417}
```

FLOOR

The FLOOR function returns the largest integer less than or equal to the specified numeric expression.

```
FLOOR(<numeric_expression>)
```

Arguments

Any expression returning numeric value.

Return types

Numeric (integer)

Example

```
SELECT { ball.num, FLOOR(ball.num / 2) AS num_val } AS num_pair
FROM blrd_store
WHERE ball.num IS NOT NULL
LIMIT 4
```

Returns

```
{
  NUM_PAIR : {
    NUM : 1,
    NUM_VAL : 0
  }
},
{
  NUM_PAIR : {
    NUM : 2,
    NUM_VAL : 1
  }
},
{
  NUM_PAIR : {
    NUM : 3,
    NUM_VAL : 1
  }
},
{
  NUM_PAIR : {
    NUM : 4,
    NUM_VAL : 2
  }
}
```

GCD

The GCD function returns the greatest common divisor of the two specified expressions.

```
GCD(<integer_expression_1>, <integer_expression_2>)
```

Arguments

Any expressions returning integer value.

If one of argument has non integer type, the function returns the error:

> ▌ `Error: wrong argument type in function GCD.`

Return type

Integer

Example

```
SELECT GCD(4, 6) AS $I;
SELECT GCD(4, -6) AS $I;
SELECT GCD(-4, 6) AS $I;
SELECT GCD(-4, -6) AS $I;
SELECT GCD(NULL, 6) AS $I;
```

Returns

```
{I:2}
{I:2}
{I:2}
{I:2}
{I:NULL}
```

LCM

The LCM function returns the least (smallest) common multiple of the two specified expressions.

```
LCM(<integer_expression_1>, <integer_expression_2>)
```

Arguments

Any expressions returning integer value.

If one of the arguments has non integer type, the function returns the error:

> Error: wrong argument type in function GCD.

Return type

Integer

Example

```
SELECT LCM(4, 6) AS $I;
SELECT LCM(4, -6) AS $I;
SELECT LCM(-4, 6) AS $I;
SELECT LCM(-4, -6) AS $I;
SELECT LCM(NULL, 6) AS $I;
```

Returns

```
{I:12}
{I:12}
{I:12}
{I:12}
{I:NULL}
```

LN

The LN function returns the natural logarithm of the specified expression.

```
LN(<numeric_expression>)
```

The natural logarithm is the logarithm to the base e, where e is an irrational constant approximately equal to 2.718281828.

Arguments

Any expression returning numeric value.

Return types

Numeric

Example

```
SELECT { ball.num, LN(ball.num) AS num_val } AS num_pair
FROM blrd_store
WHERE ball.num IS NOT NULL
LIMIT 3
```

Returns

```
{
```

```
   NUM_PAIR : {
      NUM : 1,
      NUM_VAL : 0
   }
},
{
   NUM_PAIR : {
      NUM : 2,
      NUM_VAL : 0.693147180559945
   }
},
{
   NUM_PAIR : {
      NUM : 3,
      NUM_VAL : 1.09861228866811
   }
}
```

LOG

The LOG function returns the common logarithm (the base is 10) of the specified expression.

```
LOG(<numeric_expression>)
```

Arguments

- <numeric_expression> is a numeric expression.

Return types

Numeric

Example

```
SELECT LOG(1000) AS $result
```

Returns

```
{
   RESULT : 3
}
```

MOD

The MOD function returns the remainder of the dividend number divided by the divisor number (modulus).

```
MOD(<dividend>, <divisor>)
```

Arguments

- <dividend> is an integer expression
- <divisor> is an integer expression

Return types

Numeric

Example

```
SELECT { ball.num, MOD(ball.num, 2) AS num_val } AS num_pair
FROM blrd_store
WHERE ball.num IS NOT NULL
LIMIT 3
```

Returns

```
{
  NUM_PAIR : {
    NUM : 1,
    NUM_VAL : 1
  }
},
{
  NUM_PAIR : {
    NUM : 2,
    NUM_VAL : 0
  }
},
{
  NUM_PAIR : {
    NUM : 3,
    NUM_VAL : 1
  }
},
...
```

PI

The PI function returns the constant value of pi.

```
PI()
```

Arguments

None

Return types

Numeric

Example

```
SELECT ballset.diam * PI() AS num_val
FROM blrd_store
```

Returns

```
{NUM_VAL: 193.207948195772},
{NUM_VAL: 179.542020152657}
```

POW

The POW function returns the value of the specified expression raised to the specified power.

```
POW(<base>, <exponent>)
```

Arguments

- <base> is a numeric expression.
- <exponent> is a numeric expression.

Return types

Numeric

Example

```
SELECT { ball.num, POW(ball.num, 2) AS num_val } AS num_pair
FROM blrd_store
WHERE ball.num IS NOT NULL
LIMIT 3
```

Returns

```
{
  NUM_PAIR : {
    NUM : 1,
    NUM_VAL : 1
  }
},
{
  NUM_PAIR : {
    NUM : 2,
```

```
      NUM_VAL : 4
   }
},
{
   NUM_PAIR : {
      NUM : 3,
      NUM_VAL : 9
   }
}
```

SIGN

The SIGN function returns the positive (`+1`), zero (`0`), or negative (`-1`) sign of the specified expression.

```
SIGN(<numeric_expression>)
```

Arguments

Any expression returning numeric value.

Return types

Numeric (integer). If expression is null, returns NULL.

Example

```
SELECT SIGN(5 - ball.num) AS num_val
FROM blrd_store
WHERE num = 1 OR num = 5 OR num = 6
```

Returns

```
{NUM_VAL: 1},
{NUM_VAL: 0},
{NUM_VAL: -1}
```

ROUND

The ROUND function returns a numeric value, rounded to the nearest integer value.

```
ROUND(<numeric_expression>)
```

Arguments

Any expression returning numeric value.

Return types

Numeric (integer)

Example

```
SELECT { num, ROUND(num / 3) AS num_val } AS num_pair
FROM blrd_store
WHERE ball.num IS NOT NULL
LIMIT 4
```

Returns

```
{
  NUM_PAIR : {
    NUM : 1,
    NUM_VAL : 0
  }
},
{
  NUM_PAIR : {
    NUM : 2,
    NUM_VAL : 1
  }
},
{
  NUM_PAIR : {
    NUM : 3,
    NUM_VAL : 1
  }
}
```

SIN

The SIN function returns the trigonometric sine of the specified angle in radians.

```
SIN(<numeric_expression>)
```

Arguments

Any expression returning numeric value.

Return types

Numeric

Example

```
SELECT SIN(PI() / ball.num) AS num_val, ball.num
FROM blrd_store
WHERE ball.num IS NOT NULL AND ball.num < 3
```

Returns

```
{
  RESULT : {
    #NUM_VAL : {
      NUM_VAL : 1.22464679914735e-16,
      NUM_VAL : 1
    },
    #NUM : {
      NUM : 1,
      NUM : 2
    }
  }
}
```

SQRT

The SQRT function returns the square root of the specified value.

```
SQRT(<numeric_expression>)
```

Arguments

Any expression returning numeric value.

Return types

Numeric

Example

```
SELECT { ball.num, SQRT(ball.num) AS num_val } AS num_pair
FROM blrd_store
WHERE ball.num IS NOT NULL
LIMIT 4
```

Returns

```
{
  NUM_PAIR : {
    NUM : 1,
    NUM_VAL : 1
```

```
      }
  },
  {
    NUM_PAIR : {
      NUM : 2,
      NUM_VAL : 1.4142135623731
    }
  },
  {
    NUM_PAIR : {
      NUM : 3,
      NUM_VAL : 1.73205080756888
    }
  },
  {
    NUM_PAIR : {
      NUM : 4,
      NUM_VAL : 2
    }
  }
}
```

TAN

The TAN function returns the trigonometric tangent of the specified angle in radians.

```
TAN(<numeric_expression>)
```

Arguments

Any expression returning numeric value.

Return types

Numeric

Example

```
SELECT TAN(PI() / ball.num) AS num_val
FROM blrd_store
WHERE ball.num = 2
```

Returns

```
{
  NUM_VAL : 1.63312393531954e+16
}
```

5.5   String functions

ASCII

The ASCII function returns the integer number equal to the ASCII code of the first symbol in the character string supplied as the argument.

```
ASCII(<string_expression>)
```

Arguments

Any expression returning string value.

Return types

Numeric (integer)

Example

```
SELECT ASCII('delta')
```

Returns

```
{
  RESULT : 100
}
```

CHR

The CHR function returns the character that has the ASCII code value specified by the integer argument.

```
CHR(<numeric_expression>)
```

Arguments

Any expression returning numeric value.

Return types

String

Example

```
SELECT CHR(ASCII('d')+2)
```

Returns

```
{
  RESULT : 'f'
}
```

CONCAT

The CONCAT function concatenates two strings.

```
CONCAT(<string_expression>, <string_expression>)
```

Arguments

Two expressions of string type.

Return types

String

Example

```
SELECT CONCAT('Third ball is ', ball.color) AS color
FROM blrd_store
WHERE ball.num = 3
```

Returns:

```
{
  COLOR : 'Third ball is red'
}
```

LEFT

The LEFT function extracts characters starting at the left side of specified text.

```
LEFT(<string_expression>, <count>)
```

Arguments

- <string_expression> - text to extract from.
- <count> - number of characters to extract.

Return types

String

Example

```
SELECT LEFT(ball.color, 3) AS color
FROM blrd_store
WHERE ball.num = 1
```

Returns:

```
{
  COLOR : 'yel'
}
```

LENGTH

The LENGTH function returns the number of characters of the specified string expression, including trailing spaces.

```
LENGTH(<string_expression>)
```

Arguments

Any expression returning string value.

Return types

Numeric (integer)

Example

```
SELECT LENGTH(ball.color) AS color
FROM blrd_store
WHERE ball.num = 1
```

Returns:

```
{
  NUM_VAL : 6
}
```

LOWER

The LOWER function returns a string after converting all characters to lowercase.

```
LOWER(<string_expression>)
```

Arguments

Any expression returning string value.

Return type

String

Example

```
SELECT LOWER(CONCAT('COLOR IS ', ball.color)) AS color
FROM blrd_store
WHERE ball.num = 1
```

Returns:

```
{
  COLOR : 'color is yellow'
}
```

REGEXP

The REGEXP statement implements POSIX basic regular expressions

```
<char_keyobject> REGEXP " ' " <pattern> " ' "
```

> REGEXP works in SELECT lists, in WHERE and HAVING clauses but cannot be used in JOIN conditions

Arguments

- <char_keyobject> is the keyobject or constant expression of type CHAR.
- <pattern> is a regular expression pattern to be matched.

Return types

Integer:

- 1, if pattern has matched
- 0 otherwise

Patterns

Here is the list of operators which can be used in patterns

Operator	Name	Description / Example

.	"Dot" quantifier	Matches any single character `SELECT 'KeySQL' REGEXP 'K..SQL'`
*	"Asterisk" or "star" quantifier	Matches zero or more occurrences of the subexpression/ strings preceding it `SELECT 'KeySQL' REGEXP 'Key *SQL'`
+	"Plus" quantifier	Matches one or more occurrences of the subexpression/ strings preceding it `SELECT 'KeySQL' REGEXP '.+SQL'`
?	Question mark quantifier	Matches zero or one occurrence of the subexpression/ strings preceding it `SELECT 'SQL' REGEXP '(Key)?SQL'`
[ABC] / [abc]	Matching character list	Matches any character specified in the list `SELECT 'Micro' REGEXP 'M[ai]+cro[n]*'`
[^ABC] / [^abc]	Non-matching character list	Matches any character except the ones specified in the list `SELECT 'noSQL KeySQL' REGEXP '.*[^no]SQL'`
[0-9]	Digit list	Matches any digit from 0 to 9 `SELECT 'W0rd' REGEXP 'W[0-9]rd'`
{a}	Exact count interval	Matches exact 'a' occurrences of subexpression or string preceding it `SELECT 'KeySQL' REGEXP '.{3}SQL'`

{a,}	At least one count interval	Matches at least 'a' occurrences of subexpression or string preceding it `SELECT 'KeySQL' REGEXP '.{2,}SQL'`
{a,b}	Between count interval	Matches at least 'a' occurrences of subexpression or string preceding it but not more than 'b' occurrences `SELECT 'KeySQLIsMoreThanSQL' REGEXP '.+\w{4,12}SQL'`
^	Caret quantifier	Matches an expression only if it occurs at the beginning of a line `SELECT 'KeySQL' REGEXP '^\w{3}SQL'`
$	Dollar or end quantifier	Matches an expression only if it occurs at the end of a line `SELECT 'KeySQL' REGEXP '.+SQL$'`
(|)	Vertical bar quantifier	Used to specify alternatives `SELECT 'SQL database' REGEXP '.+\s+(SQL|data).+'`

[[:class:]]	Class quantifier	Matches the character class:
		<ul><li>[[:digit:]] for digits,</li><li>[[:space:]] for whitespaces,</li><li>[[:alnum:]_] for alphanumeric (digits and letters),</li><li>etc.</li></ul> `SELECT '123 ABC' REGEXP '\d{3}\s{1}\w{3}'` Class-shorthand Escapes <ul><li>\d - [[:digit:]]</li><li>\s - [[:space:]]</li><li>\w - [[:alnum:]_] (underscore is included)</li><li>\D - [^[:digit:]]</li><li>\S - [^[:space:]]</li><li>\W - [^[:alnum:]_] (underscore is included)</li></ul>

Examples

Example 1

```
SELECT game
FROM blrd_store
WHERE ball.color REGEXP '(white|red)+'
```

Result

```
{ GAME : 'carom' },
{ GAME : 'pool' }
```

Example 2

```
SELECT { game, color } AS $colors
FROM blrd_store
WHERE ball.color REGEXP '[^oe]+'
```

Result

```
{
    COLORS : { GAME : 'pool', COLOR : 'pink' }
},
```

```
{
  COLORS : { GAME : 'pool', COLOR : 'tan' }
},
{
  COLORS : { GAME : 'pool', COLOR : 'black' }
}
```

REPEAT

The REPEAT function repeats a string value a specified number of times.

```
REPEAT(<string_expression>, <count>)
```

Arguments

- <string_expression> - string to replicate.
- <count> - number of repetitions

Return types

String

Example

```
SELECT REPEAT(ball.color,3) AS $blablabla
FROM blrd_store
WHERE ball.num = 2
```

Returns

```
{
  BLABLABLA : 'blueblueblue'
}
```

REPLACE

The REPLACE function replaces all occurrences of a specified string value with another string value.

```
REPLACE(<string_expression1>, <string_expression2>,
<string_expression3>)
```

Arguments

- <string_expression1> - string to search in
- <string_expression2> - string to find in <string_expression1>
- <string_expression3> - string to replace with

Return types

String

Example

```
SELECT REPLACE(ball.color, 'ow', ' on') AS color
FROM blrd_store
WHERE ball.num = 1
```

Returns

```
{
  COLOR : 'yell on'
}
```

REVERSE

The REVERSE function returns the reverse ordered string.

```
REVERSE(<string_expression>)
```

Arguments

Any expression returning string value

Return types

String

Example

```
SELECT REVERSE(ball.color) AS $roloc
FROM blrd_store
WHERE ball.num = 1
```

Returns

```
{
  ROLOC : ' wolley'
}
```

RIGHT

The RIGHT function returns the right part of a string with the specified number of characters.

```
RIGHT(<string_expression>, <count>)
```

Arguments

- <string_expression> - string to extract from.
- <count> - number of rightmost characters to extract.

Return types

String

Example

```
SELECT RIGHT(ball.color, 3) AS color
FROM blrd_store
WHERE ball.num = 1
```

Returns:

```
{
   COLOR : 'low'
}
```

STRPOS

The STRPOS function searches for a string inside the second string, returning the starting position of the first one if found.

```
STRPOS(<string_expression1>, < string_expression2>)
```

Arguments

- <string_expression1> - string to search for.
- <string_expression2> - string to search in

Return types

Integer. Character position in the string starts from 1.

Returns 0 when not found.

Example

```
SELECT STRPOS(ball.color, 'low') AS num_val
FROM blrd_store
WHERE ball.num = 1
```

Returns:

```
{ NUM_VAL : 4 }
```

SUBSTRING

The SUBSTRING function returns a part of a string.

```
SUBSTRING(<string_expression>, <position>, <count>)
```

Arguments

- <string_expression> - string to extract from.
- <position> - starting position (from 1).
- <count> - number of characters to extract.

Return types

String

Example

```
SELECT SUBSTRING(ball.color, 4, 3) AS color
FROM blrd_store
WHERE ball.num = 1
```

Returns:

```
{
   COLOR : 'low'
}
```

UPPER

The UPPER function returns a string after converting all characters to uppercase.

```
UPPER(<string_expression>)
```

Arguments

Any expression returning string value.

Return types

String.

Example

```
SELECT UPPER(ball.color) AS color
FROM blrd_store
WHERE ball.num = 1
```

Returns

```
{
   COLOR : ' YELLOW'
}
```

5.6   Date and time functions

Consider creating the following test data to run examples of datetime functions.

```
CREATE CATALOG date_functions;
CREATE KEYOBJECT date_val DATE IN CATALOG date_functions;
CREATE KEYOBJECT str_val CHAR IN CATALOG date_functions;
CREATE KEYOBJECT num_val NUMBER IN CATALOG date_functions;
CREATE STORE date_test FOR CATALOG date_functions;
INSERT INTO date_test INSTANCES
{date_val: '2021-01-01'},
{date_val: '2021-01-02 12:34:56'},
{date_val: '2021-01-02 12:34:56.78901'};
```

CURRENT_DATE

The CURRENT_DATE function returns the current database system timestamp without the time zone offset.

```
CURRENT_DATE()
```

Arguments

None

Return types

Date (timestamp without time zone)

Example

```
SELECT CURRENT_DATE() AS $D
```

Returns:

```
{
   D :  '2022-11-26 16:00:17.867153'
}
```

EXTRACT

The EXTRACT function returns the part of the date value.

```
EXTRACT(<date_part_name> FROM <date_value>)
```

Arguments

- `<date_part_name>` is one of supported date parts:
  - year
  - month
  - day
  - hour
  - second
  - microsecond
- `<date_value>` - any value of DATE type

Return types

Integer.

Example

```
SELECT EXTRACT(year FROM date_val) AS num_val
FROM date_test
```

Returns

```
{ NUM_VAL : 2021 },
{ NUM_VAL : 2021 },
{ NUM_VAL : 2021 }
```

5.7   System functions

CAST

The CAST function converts an expression of one data type to another. See "Elementary data types".

```
CAST(<expression> AS <elementary_data_type>)
```

Arguments

- `<expression>` - expression returning an elementary k-object.

- `<elementary_data_type>` - one of the supported elementary data types.

The conversion compatibility table is as follows.

	NUMBER	INTEGER	CHAR	DATE
NUMBER	-	YES	YES	NO
INTEGER	YES	-	YES	NO
CHAR	MAYBE	MAYBE	-	MAYBE
DATE	NO	NO	YES	-

Where:

- YES - conversion is allowed.
- NO - conversion is not allowed.
- MAYBE - conversion depends on the value.

For example, an attempt to convert an alpha string to integer value will produce an error.

```
SELECT CAST('ABC' AS INTEGER)
```

> Error: invalid CAST input for type INTEGER: 'ABC'

However, a "numeric" string will be converted correctly:

```
SELECT CAST('123' AS INTEGER)
```

Result:

```
{
  RESULT : 123
}
```

The following date string conversion

```
SELECT CAST('2020-01-01' AS DATE)
```

works as expected, too:

```
{
```

```
    RESULT : '2020-01-01'
}
```

Return types

Specified by `<elementary_data_type>`.

When the type conversion is allowed, the converted value is returned; otherwise, the statement produces an error.

Examples

Create test catalog and store.

```
CREATE CATALOG cast_test;

CREATE KEYOBJECT num_val NUMBER IN CATALOG cast_test;
CREATE KEYOBJECT str_val CHAR IN CATALOG cast_test;
CREATE KEYOBJECT str_val2 CHAR IN CATALOG cast_test;
CREATE KEYOBJECT date_val DATE IN CATALOG cast_test;
CREATE KEYOBJECT test {num_val, str_val, str_val2, date_val} IN
CATALOG cast_test;
CREATE STORE cast_test_store FOR CATALOG cast_test;
INSERT INTO cast_test_store INSTANCES
{test: {num_val: 123, str_val: '456', str_val2: '2020-02-02',
date_val: '2021-01-01'}}
```

Consider two exemplary cast queries.

```
SELECT
    CAST(num_val AS CHAR) AS str_val,
    CAST(str_val AS NUMBER) AS num_val,
    CAST(str_val2 AS DATE) AS date_val,
    CAST(date_val AS CHAR) AS str_val2
FROM cast_test_store;
```

This statement produces the following result.

```
{
  RESULT : {
    STR_VAL : '123',
    NUM_VAL : 456,
    DATE_VAL : '2020-02-02',
    STR_VAL2 : '2021-01-01 00:00:00'
  }
}
```

The next statement produces an error as follows.

```
SELECT
    CAST(num_val AS DATE) AS date_val
FROM cast_test_store;

    Error: invalid CAST input
```

KeySQL System k-objects 6

When inserted into a store, each host instance is complemented with two default system attributes, which are the instances of the reserved k-objects that are not present in any catalog.

_IID

The `_IID` represents a unique (within the store) and permanent host instance identifier.

`_IID INTEGER`

The values of `_IID` can be retrieved like the values of catalog k-objects with some limitations. Refer to SELECT statement for more details.

Example

```
SELECT _IID, ball FROM my_store
```

_VERSION

The `_VERSION` is an incremental version number of the host instance – any update of an instance increments its version by 1.

`_VERSION INTEGER`

The values `_VERSION` can be retrieved like the values of catalog k-objects with some limitations. Refer to SELECT statement for more details.

Example

```
SELECT ball, _VERSION FROM my_store
```

Index

www.ingramcontent.com/pod-product-compliance
Lightning Source LLC
LaVergne TN
LVHW080106070326
832902LV00034B/2217